# Environmental Politics and Theory

Series Editors
Joel Jay Kassiola, Department of Political Science, San Francisco State
University, San Francisco, CA, USA
John Barry, School of History, Anthropology, Philosophy and Politics,
Queen's University Belfast, Belfast, UK

The premise of this series is that the current environmental crisis cannot be solved by technological innovation alone and that the environmental challenges we face today are, at their root, political crises involving political values. Therefore, environmental politics and theory are of the utmost social significance. Growing public consciousness of the environmental crisis and its human and nonhuman impacts exemplified by the worldwide urgency and political activity associated with the problem and consequences of climate change make it imperative to design and achieve a sustainable and socially just society. The series collects, extends, and develops ideas from the burgeoning empirical and normative scholarship spanning many disciplines with a global perspective. It addresses the need for social change from the hegemonic consumer capitalist society in order to realize environmental sustainability and social justice.

More information about this series at
http://www.palgrave.com/gp/series/14968

John S. Duffield

# Making Renewable Electricity Policy in Spain

The Politics of Power

John S. Duffield
Department of Political Science
Georgia State University
Atlanta, GA, USA

Environmental Politics and Theory
ISBN 978-3-030-75640-6     ISBN 978-3-030-75641-3  (eBook)
https://doi.org/10.1007/978-3-030-75641-3

© The Editor(s) (if applicable) and The Author(s), under exclusive license to Springer Nature Switzerland AG 2021

This work is subject to copyright. All rights are solely and exclusively licensed by the Publisher, whether the whole or part of the material is concerned, specifically the rights of translation, reprinting, reuse of illustrations, recitation, broadcasting, reproduction on microfilms or in any other physical way, and transmission or information storage and retrieval, electronic adaptation, computer software, or by similar or dissimilar methodology now known or hereafter developed.

The use of general descriptive names, registered names, trademarks, service marks, etc. in this publication does not imply, even in the absence of a specific statement, that such names are exempt from the relevant protective laws and regulations and therefore free for general use.

The publisher, the authors and the editors are safe to assume that the advice and information in this book are believed to be true and accurate at the date of publication. Neither the publisher nor the authors or the editors give a warranty, expressed or implied, with respect to the material contained herein or for any errors or omissions that may have been made. The publisher remains neutral with regard to jurisdictional claims in published maps and institutional affiliations.

Cover illustration: Evandro Maroni/Stockimo/Alamy Stock Photo

This Palgrave Macmillan imprint is published by the registered company Springer Nature Switzerland AG
The registered company address is: Gewerbestrasse 11, 6330 Cham, Switzerland

*For Elizabeth and Stuart*

# Series Editors' Preface for: John S. Duffield

## Introduction

We are delighted and proud to welcome John S. Duffield and his volume, *The Politics of Power: Making Renewable Electricity Policy in Spain*, to the Palgrave Macmillan Environmental Politics and Theory Series (EPT).

Duffield's analysis of the politics of Spanish renewable electricity policy continues a distinctive feature of the EPT Series going back to the first publication within the Series: single-country case studies on the environmental problems and policy responses to them. Beginning with *China's Environmental Crisis: Domestic and Global Political Impacts and Responses* (Kassiola and Guo), continuing with a book on the environmental conditions and politics in Mexico, *Social Environmental Conditions in Mexico* (Tetreault, McCulligh, and Lucio), then a volume on China's governance of its environment and responses to environmental challenges, *Chinese Environmental Governance: Dynamics, Challenges and Prospects in a Changing Society* (Ren and Shou), the EPT Series has recognized the important contributions that national case studies of a country's environmental conditions and its policy responses make in increasing our understanding and addressing our environmental predicament.

With the publication of *The Politics of Power*, we aspire to make the EPT Series the home for a new scholarly subfield of comparative environmental politics. It focuses on the environmental circumstances, issues and governmental responses of various nations in order to accumulate informative data and analyses.

## The Structure of the Duffield Volume and Its Theme

No current environmental problem is more urgent than global warming and climate change. Therefore, no policy response to this dire problem is more pressing than replacing fossil fuel-generated power with renewable and non-greenhouse gas-emitting sources of power. Duffield's account in this book subtitled "making renewable electricity policy in Spain" is both comprehensive and historical. It describes and explains renewable electricity generation and policy in Spain from its origins in the 1970s to the present and beyond with projections to 2030. *The Politics of Power* is, therefore, a timely and important study given Spain's worldwide leadership in renewable electricity sources: wind power, solar photovoltaic (PV), and concentrated solar power (solar thermal).

By learning the development of Spanish renewable electricity policy—which was not a straight line upward, as Duffield reports—comparative political scientists, energy specialists, and energy policy-makers across the globe as well as students of Spanish politics will be able to gain important insight. All could then make significant contributions to their nations' renewable energy policy and its implementation: a goal vital to reducing consequences of the harmful climate change.

Duffield says that his study raises the following central question: "What accounts for Spain's early leadership role and the substantial fluctuations in support and growth that have subsequently marked its relationship with renewable power?" (Chapter 1). His reply to this query is to examine the politics of renewable power in Spain by exploring the evolution of Spanish renewable energy policy from the decades after the adoption of the Constitution of 1978 when Spain made its first modest efforts at renewable power (Chapter 2), followed by the so-called classical period of renewable energy policy (Chapter 3), supported by the bi-partisan "special regime" when Spain experienced rapid growth in wind power and solar PV as well as concentrated solar power. This classical period was succeeded by the "dark ages," according to Duffield, from the late 2000s to the mid-2010s, where both Spanish major parties agreed to terminate the special policy regime of providing national government financial support for renewable energy sources that resulted in ceasing new construction of renewable installations and plant owners threatened with bankruptcy (Chapter 4). The ups and downs of Spanish renewable power policy are highlighted again, according to Duffield's narrative, when in the late 2010s there was a "renaissance" in renewable power construction of new renewable power plants with a doubling of the amount of solar PV (Chapter 5).

Possibly the most significant aspect of the evolution of Spanish renewable power generation for comparativists and policy-makers alike is the development of what is termed "self-consumption," or the distribution of "renewable electivity generation that is largely consumed on site by those who produce it." This mode of energy production and consumption expanded rapidly under the current Socialist government beginning in the 2010s (Chapter 6), and holds lessons for other countries' efforts to create decentralized renewable power generation.

Duffield's study is not only valuable for its analyses of the current status of Spanish renewable electricity and policy but in its projecting future prospects for these modes of non-fossil fuel sources of power to the year 2030. The author singles out 3 challenges facing Spain in the next 10 years and indeed these issues are global in scope and application regarding the scaling up of renewable electricity so necessary to reduce our current greenhouse emissions from fossil fuel-based energy sources: coal, natural gas, and petroleum. The challenges are: (a) building sufficient generating capacity; (b) accommodating the increased renewable power into the electric power grid; and (c) incorporating the increased renewable power generated into the existing electricity market (Chapter 7).

Finally, Chapter 8 provides an explanatory framework for the previous empirical chapters and should be of most generalizable theoretical interest. Here, Duffield seeks to identify the most important actors and institutional structures in the renewable power policy-making process. He finds the key actor in this process to be the central executive branch of the national government. He also reports that the political party in power does not always follow their party's ideology regarding renewable electricity.

Interestingly, Duffield identifies the importance of external forces upon Spain's political system and renewable power policy, primarily the EU and its requirements impacting social and economic factors essential to the construction and generation of renewable electricity power plants. Rounding out the key actors in the Spanish renewable energy policy story are various interest groups within Spain; a development that should come as no surprise to political scientists interested in the environment (Chapter 8).

## Conclusion

The reader is about to embark upon a historical and political journey through four decades of Spanish renewable power policy-making marked by fundamental shifts of government support. For an analogy to the American context, we may think of the various shifts on climate change inherent in the Bush-Obama-Trump-Biden presidencies over the past 20 years. The goal for environmentalists must be to achieve sustained and consistent bi-partisan support for renewable energy. The world cannot tolerate any more backsliding given the pressure of time to reduce our greenhouse emissions. What can we learn from the Spanish experience? That is the contribution of comparative studies such as Duffield's.

The increased use renewable electricity and the cessation of reliance upon fossil fuels are ecological imperatives that must be integrated into various national policy-making systems to become political imperatives worldwide. We think Duffield's study will both inform and help provide a structure for that political change. We hope that the readers of *The Politics of Power*, especially decision-makers, will agree.

Joel Jay Kassiola
John Barry
EPT Series Co-Editors

# Acknowledgments

A number of individuals and institutions contributed to the completion of this book. For financial support, I am grateful to the College of Arts and Sciences and the Department of Political Science at Georgia State University, which provided funding that enabled me to make two research trips to Spain. I also wish to thank the many Spanish experts in government, academia, research centers, industry associations, and advocacy groups who have helped me to understand this fascinating but often complex subject. Nevertheless, I alone am responsible for any errors of fact or interpretation that remain.

As always, my wife Cheryl has been a constant source of support throughout the process of researching and writing this book. Without her encouragement, I never would have finished it.

Finally, I dedicate the book to my children. It was they who introduced me to Spain and helped to kindle my passion for its people, history, and culture. That I was able to connect that passion to my research agenda was an unanticipated bonus. For all the ways in which you now teach me, thank you, kids!

# Contents

| | | |
|---|---|---|
| 1 | **Introduction** | 1 |
| | *The Puzzles of Renewable Power Policy in Spain* | 1 |
| | *Contributions to the Literature* | 4 |
| | *Analytic Framework and Expectations* | 5 |
| | *Central Arguments* | 7 |
| | *Organization* | 8 |
| | *References* | 10 |
| 2 | **Laying the Groundwork: Early Efforts to Promote Renewable Power in Spain** | 13 |
| | *Introduction* | 13 |
| | *The Legacy of Franco* | 13 |
| | *The 1980s: Initial Efforts to Promote Renewable Power* | 14 |
| | *The 1990s: The First Systematic Support for Renewable Power* | 18 |
| | *Conclusion* | 21 |
| | *References* | 22 |
| 3 | **The Classical Period: Renewable Power Takes Off** | 25 |
| | *Introduction* | 25 |
| | *First Steps: The New "Special Regime"* | 26 |
| | *Initial Adjustments Under the Aznar Government* | 32 |
| | *Further Adjustments Under the Socialists:* Real Decreto *661/2007* | 37 |

| | |
|---|---:|
| *Conclusion* | 45 |
| *References* | 47 |

## 4 The Dark Ages: Reponses to the Renewables Boom    51

| | |
|---|---:|
| *Introduction* | 51 |
| *Motivations for Reform: Negative Consequences of the Boom* | 52 |
| *Efforts to Limit the Growth in Support Costs: To the Moratorium* | 57 |
| *Efforts to Reduce Existing Costs* | 61 |
| *The End of the Special Regime* | 66 |
| *Long-Term Consequences of the Reforms* | 69 |
| *Conclusion* | 70 |
| *References* | 71 |

## 5 The Renaissance of Renewable Power    77

| | |
|---|---:|
| *Introduction* | 77 |
| *Pressures for Further Government Action* | 78 |
| *The Rajoy Government's Response: The Auctions* | 81 |
| *Market Developments* | 86 |
| *Impact on Deployment* | 89 |
| *Conclusion* | 91 |
| *References* | 92 |

## 6 The Battle Over Self-Consumption    101

| | |
|---|---:|
| *Introduction* | 101 |
| *Why Self-Consumption?* | 102 |
| *Initial Efforts to Promote and Regulate Self-Consumption* | 103 |
| *The Rajoy Government's Counterreformation* | 107 |
| *Political Counterpressure: Efforts to Reverse* Real Decreto *900/2015* | 113 |
| *Reversal of Fortune Under the Sánchez Government* | 116 |
| *Conclusion* | 122 |
| *References* | 124 |

## 7 Future Prospects for Renewable Power in Spain    135

| | |
|---|---:|
| *Introduction* | 135 |
| *Motivations* | 136 |
| *Renewable Energy Targets* | 136 |
| *Can Sufficient Capacity Be Built?* | 140 |
| *Can so Much Capacity Be Accommodated by the Electrical Grids?* | 147 |

|     |                                                                         |     |
| --- | ----------------------------------------------------------------------- | --- |
|     | *Can the New Renewable Capacity Meet Demand?*                           | 152 |
|     | *Conclusion*                                                            | 155 |
|     | *References*                                                            | 156 |
| 8   | **The Politics of Renewable Power in Spain**                            | 163 |
|     | *Introduction*                                                          | 163 |
|     | *The Primacy of the Executive*                                          | 164 |
|     | *Sources of Convergent Policy Preferences*                              | 166 |
|     | *Sources of Divergent Policy Preferences*                               | 170 |
|     | *Other National Actors*                                                 | 173 |
|     | *The Regions and Municipalities*                                        | 175 |
|     | *Conclusion: The Future of Renewable Power Policymaking in Spain*       | 180 |
|     | *References*                                                            | 181 |
|     | **Index**                                                               | 185 |

# About the Author

**John S. Duffield** is Professor of Political Science and Director of Assessment and Review at Georgia State University in Atlanta. He received a doctorate in public and international affairs from Princeton University. His teaching and research focus on international politics and the politics of energy and climate change, both in the United States and abroad. He is the author of four previous books and has co-edited two others: *Fuels Paradise: Seeking Energy Security in Europe, Japan, and the United States* (Johns Hopkins University Press, 2015); *Over a Barrel: The Costs of U.S. Foreign Oil Dependence* (Stanford University Press, 2008); *World Power Forsaken: Political Culture, International Institutions, and German Security Policy after Unification* (Stanford University Press, 1998); *Power Rules: The Evolution of NATO's Conventional Force Posture* (Stanford University Press, 1995); Vicki L. Birchfield and John S. Duffield, eds., *Toward a Common EU Energy Policy: Progress, Problems, and Prospects* (Palgrave Macmillan, 2011); and John S. Duffield and Peter J. Dombrowski, eds., *Balance Sheet: The Iraq War and U.S. National Security* (Stanford University Press, 2009).

# Acronyms

| | |
|---|---|
| Anpier | *Asociación Nacional de Productores de Energía Fotovoltaica* |
| APPA | *Asociación de Empresas de Energías Renovables* |
| ASIF | *Asociación de la Industria Fotovoltaica* |
| CCAA | *comunidades autónomas* |
| CNC | *Comisión Nacional de la Competencia* |
| CNE | *Comisión Nacional de la Energía* |
| CNMC | *Comisión Nacional de los Mercados y la Competencia* |
| CPI | Consumer Price Index |
| CSP | Concentrated Solar Power |
| EC | European Commission |
| EU | European Union |
| GFCC | Gas-Fired Combined Cycle |
| GW | Gigawatt (1 billion watts) |
| GWh | Gigawatt-hour |
| IDAE | *Instituto para la Diversificación y Ahorro de la Energía* |
| IEA | International Energy Agency |
| IRENA | International Renewable Energy Association |
| kW | kilowatt (1 thousand watts) |
| kWh | kilowatt-hour |
| LCCTE | *Ley de Cambio Climático y Transición Energética* |
| LCOE | Levelized Cost of Electricity |
| MINETUR | *Ministerio de Industria, Energía y Turismo* |
| MITECO | *Ministerio para la Transición Ecológica* |
| MW | Megawatt (1 million watts) |
| MWh | Megawatt-hour |
| OECD | Organization for Economic Cooperation and Development |

| | |
|---|---|
| PEN | *Plan Energético Nacional* |
| PER | *Plan de Energías Renovables* |
| PNA | *Plan Nacional de Asignación de Derechos de Emisión* |
| PNIEC | *Plan Nacional Integrado de Energía y Clima* |
| PP | *Partido Popular* |
| PPA | Power Purchase Agreement |
| PSOE | *Partido Socialista Obrero Español* |
| PV | Photovoltaic |
| REE | *Red Eléctrica de España* |
| REER | *Régimen Económico de Energías Renovables* |
| TEM | *Tarifa eléctrica media* |
| TWh | Terawatt-hour |
| UCD | *Unión de Centro Democrático* |
| UNEF | *Unión Española Fotovoltaica* |

# List of Figures

| | | |
|---|---|---|
| Fig. 1.1 | Renewable generating capacity, 1998–2020 (MW) (*Source* CNMC 2021, Table 1.2) | 3 |
| Fig. 1.2 | Renewable power sold, 1998–2020 (GWh) (*Source* CNMC 2021, Table 1.2) | 4 |
| Fig. 4.1 | Tariff deficit, 2000–2013 (millions of euros) (*Source* Monforte 2020b) | 54 |
| Fig. 6.1 | New self-consumption generating capacity (MW) (*Sources* UNEF 2020, 79, and Roca 2021) | 122 |

# List of Tables

| | | |
|---|---|---|
| Table 3.1 | Renewable power capacity and targets, 1998–2010 (MW) | 30 |
| Table 3.2 | Renewable power generation and targets, 1998–2010 (GWh) | 30 |
| Table 5.1 | Renewable power targets for 2020 | 80 |
| Table 5.2 | Auction results, 2016–2017 | 83 |
| Table 7.1 | PNIEC capacity targets (MW) | 139 |
| Table 7.2 | Auction schedule (minimum amounts of capacity in MW) | 147 |
| Table 8.1 | Energy plans and promotional measures in Andalusia | 177 |

CHAPTER 1

# Introduction

## THE PUZZLES OF RENEWABLE POWER POLICY IN SPAIN

Climate change has become one of the defining issues of our time. Evidence that the climate is undergoing a profound transformation continues to mount, often faster than scientists had expected. And the consequences of unchecked climate change could be disastrous for much of the world's population. Even wealthy countries such as the United States face potential impacts that will be difficult to manage.

Not surprisingly, given the level of concern about and growing repercussions of climate change, there is no shortage of proposals for how to respond to this challenge. Some proposals fall under the heading of adaptation, given the inevitability of some degree of climate change. Indeed, many adaptation efforts are already underway. But most observers agree that at least some—and most likely substantial—effort is needed to mitigate the effects of climate change, given how devastating its impact could be. Here, proposals range widely, from modifying the earth's atmosphere to reflect more of the sun's radiation, to removing substantial amounts of climate change causing greenhouse gases from the atmosphere, and finally, to reducing the emission of those gases in the first place.

Perhaps the most promising and least controversial approaches fall into the latter category, and one area that has received particular attention is the replacement of fossil fuels with renewable sources of energy. The burning of fossil fuels for electricity, heat, and transportation is the single largest source of greenhouse gas emissions from human activity. Yet most if not all of those uses can be met by energy from other sources, including renewable ones. And of those, the renewable source with the greatest range of applications is renewable electricity.

In fact, power generated by renewable sources, but especially hydroelectric, wind, and solar photovoltaic (PV) power, has been growing at an accelerating pace. From 2009 to 2019, global renewable electricity output jumped from just under 3900 terawatt-hours (TWh) to more than 7000 TWh, and the rate of annual increase has steadily risen. Over the same period, the share of all renewable energy consumed in the form of electricity advanced from 72.9 to 86.3%, reflecting renewable power's greater versatility (BP 2020). And in recent years, the lion's share of that growth has been in wind and PV.

Nevertheless, there has been and continues to be considerable variation in the degree to which individual countries produce and consume renewable power. Even among the advanced industrialized countries of the Organization for Economic Cooperation and Development, per capita renewable generation in 2019 ranged from as low a 1 megawatt-hour (MWh) in Hungary to a high of 142 MWh in Iceland (Ritchie and Roser 2020). Some of this variation may be attributable to differences in national resource endowments, such as rainfall, sunshine, and wind. But other causes are political, economic, and sociological in nature—and thus more subject to human control.

What are the prospects for rapidly increasing the production and use of renewable electricity, especially wherever the availability of natural resources makes doing so possible? There are many ways to answer this question, but one promising place to begin is by examining the experience of countries that have already made major efforts to promote renewable power. What methods have they used? How successful have they been? What obstacles have they encountered? How have obstacles been overcome? What other, sometimes unintended, consequences have these efforts had? And, above all, why have they done so?

In this regard, Spain provides a potentially instructive example. Beginning in the 1990s, Spain was one of the first countries to strongly promote renewable power. As a result of these efforts, renewable generating capacity and electricity output increased substantially during the first decade of the twenty-first century, and Spain became a world leader in wind, PV, and concentrated solar power (CSP, also known as solar thermal). In the last few years, moreover, the country has experienced a renewed burst of renewable construction comparable in magnitude to that of the original buildup, and it has set some of the most ambitious targets for renewable power in the world (Figs. 1.1 and 1.2).

Spain's renewable power story has not been one of uninterrupted growth, however. The initial boom was followed by sharp cuts in the amount of support awarded to new projects, then a moratorium on additional assistance, and, finally, reductions in the level of support already promised to existing renewable power plants. For a handful of years, moreover, Spanish policy made it unprofitable for consumers, both large and small, to generate their own renewable electricity. Only in the midto late 2010s did the Spanish government once again begin to promote the deployment of additional renewable capacity, and then only when the cost of renewables had dropped to a level where public financial support was less critical to their success.

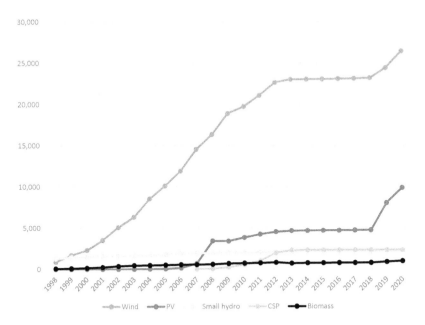

**Fig. 1.1** Renewable generating capacity, 1998–2020 (MW) (*Source* CNMC 2021, Table 1.2)

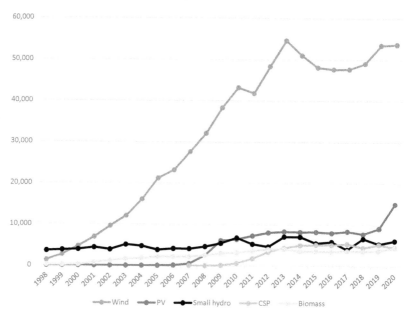

**Fig. 1.2** Renewable power sold, 1998–2020 (GWh) (*Source* CNMC 2021, Table 1.2)

## Contributions to the Literature

What accounts for Spain's early leadership role and the substantial fluctuations in support and growth that have subsequently marked its relationship with renewable electricity? This project seeks to answer this central question through an examination of the politics of renewable power in Spain. It has two overall objectives. The first is to describe in detail the important actions Spanish leaders have taken with regard to renewable power over the years, from the restoration of democracy to the present, and the consequences of those actions. The second is to provide a comprehensive explanation of the evolution of Spanish policy. Which political actors have mattered the most, and why did they act as they did?

To be sure, a considerable amount has already been written on renewable power in Spain. Of particular note are the numerous works by Pablo del Río and Pere Mir-Artigues, either individually, jointly, or with others (e.g., del Río 2016; del Río 2017; del Río and Gaul 2007; del Río and

Mir-Artigues 2012; del Río and Mir-Artigues 2014; del Río González 2008; Mir-Artigues 2012; Mir-Artigues 2013; Mir-Artigues et al. 2018; Sáenz de Miera et al. 2008). Collectively, these works touch upon many aspects of the subject, and they are cited with some frequency in the following chapters.

Nevertheless, this book stands out from the existing literature in several respects. One is its substantive scope. The book addresses all aspects of renewable power policy. Many works examine policy toward just one technology or another, such as wind power or solar power, or just one modality of generation or another, either utility-scale or small distributed. As a result, they may miss important relationships, synergies, and tradeoffs across the different modalities and technological realms.

A second is the book's chronological scope. It embraces the entire history of renewable power policy in democratic Spain, from its origins to the present. Most existing works examine only a limited time period and a temporally restricted set of policies. As a result, they may overlook important patterns in how policies have changed or remained the same over time.

The third and perhaps most important difference is the book's analytical focus. Most existing works take a primarily economic approach. They seek to evaluate the effectiveness or cost-effectiveness of different policies and often offer prescriptions for improving them. The focus of this book is primarily political. It seeks to understand how and why particular policies have been adopted in the first place, given the range of options that political leaders have had at their disposal. In short, the book offers the first comprehensive analysis of the politics of renewable power in Spain.

A final factor is the book's timeliness. Much of the existing literature concerns Spain's initial renewable power boom and bust of the first decade and a half of the century. This project brings the analysis up to date, covering the revival of interest in renewable power in Spain over the last several years and reviewing its future prospects.

## Analytic Framework and Expectations

Where should we look for the principal political determinants of renewable power policy in Spain? The broader literature on public policy in democratic Spain suggests the value of a three-level framework of analysis. At one level is the European Union (EU). Since—and even before—Spain joined the EU in 1986, the regional body has been an important source of opportunities, imperatives, and constraints that have shaped Spanish policies. With regard to renewable energy in particular, the EU has set a series of collective and individual targets that Spain has had to comply with as well as, increasingly, rules governing the allowable means of doing so. Nevertheless, Spain has generally enjoyed considerable autonomy in determining how to meet those targets, as well as the freedom to set even more ambitious objectives (Heywood 1995, 98, 242).

The next, and arguably the most important, level is the national political arena. And, here, the most significant actor has traditionally been viewed as the core executive of the central government. As a general rule, political power has been highly concentrated, and the core executive has dominated the policy process and thereby exerted the greatest amount of influence over national laws, policies, and regulations. In contrast, the Spanish parliament has been widely seen as a weak institution, one that has been incapable of exercising any effective control over the functioning of the executive (Heywood 1995, 241–248; 1999, 103–106, 114; Chari and Heywood 2009).

Within the central government, the leading role has been played by the President, who is the unambiguous head of the core executive. As Heywood has noted, the President is constitutionally authorized to monopolize the most important decisions over national policy. The President appoints the other members of the cabinet, and whenever there is disagreement within the cabinet, the President is empowered to take the final decision (Heywood 1995, 88–91; 1999, 106).

What remains to be determined are the sources of the interests that the central government seeks to pursue and promote. An obvious starting point is the ideology of the political party that happens to be in power at a particular time. Traditionally, control of the reins of government has alternated between the center-right People's Party (*Partido Popular* or PP) and the center-left Spanish Socialist Workers' Party (*Partido Socialista Obrero Español* or PSOE). Over the last decade, however, the situation has grown more complicated with the emergence of several more parties with national appeal, the left-wing *Podemos* (We Can), the liberal (in the European sense) *Ciudadanos* (Citizens), and the far-right *Vox*. As a result, the PP and PSOE have had to form minority governments reliant on outside support and, most recently, a true coalition government. In any case, we might expect changes in policy with changes in the ruling party or parties in power.

Another potential source of the government's policy preferences is interest groups. Traditionally, interest groups in Spain have been regarded as weak and of limited influence. Compared with those in other democracies, they have had substantially less access to the core executive, parliament, political parties, and politicians in general. Their role has been largely limited to reacting to, rather than driving, policy initiatives. One exception to this tendency, however, may be highly regulated areas, such as energy. There, it is much more likely that interest groups with the power to influence government policy exist (Medina et al. 2016). Moreover, it may be difficult to assess the influence of such groups, as what relations they have had with the central government have tended to be shrouded in secrecy, occurring largely through private and informal channels (Molins and Casademunt 1998).

In the Spanish case, there is also a third political level to consider. Although Spain is not a federal state, it is highly decentralized, consisting of 17 regions or autonomous communities (*comunidades autónomas* or CCAA) and two autonomous cities. The governments of these regions hold a notable amount of political power, and this authority has expanded over time. What impact the regional level will have on renewable power will depend critically on what precise competencies and resources the regions possess (Heywood 1995, 97–98).

## Central Arguments

Overall, the book finds that the core executive has indeed been the most important actor in the process of determining renewable power policy. Although the EU has provided targets, it has generally found a willing partner in the Spanish government, which has sometimes even sought to exceed the minimum requirements. Nor has the executive been significantly constrained by the national legislature. When the party in power has enjoyed a majority in the Congress, it has had no trouble passing any desired legislation. And on those occasions when it has lacked a reliable majority for major initiatives, the executive has nevertheless been able to amend existing laws and approve regulations largely as needed.

The book also finds, however, that changes in policy, and the resulting patterns of deployment, have not primarily been the result of variations in the parties in power. To the contrary, as others have noted, the major parties have generally held similar views on the desirability of promoting renewable power and pursued similar policies. Support for renewable power has been a widely shared priority, reflecting both Spain's resource endowments and the potential security, economic, and environmental benefits of renewables (del Río and Gaul, 2007; del Río 2008; Linares and Labandeira 2013). What has been more determinative of policy change, especially reductions in the level of support, has been variation in Spain's economic and financial circumstances and thus the country's ability to provide support to renewable power, as well as changes in the cost of building new renewable power plants.

This broad consensus has not extended to all areas of renewable power policy, however. The most striking exception is offered by Spanish policy on electricity that is self-generated and largely or entirely self-consumed, or what is commonly called self-consumption (*autoconsumo*). This area has been characterized by heated political disagreement that stymied its development for a number of years. Where such differences in policy preferences do occur, they can be attributed largely to differences in party ideology, variations in the influence of different interest groups, and even the personalities of key individual leaders.

## Organization

The remainder of the book consists of seven chapters. The next five detail how Spain's renewable power policies have evolved and with what political consequences. Chapter 2 reviews developments during the first two decades after the adoption of the new democratic Constitution in 1978. During those early years, Spain took its first steps toward promoting renewable power. Although the actual level of production remained modest, governments of both the center-right and the center-left laid the groundwork for the major growth that was to follow.

Chapter 3 covers the critical decade from the late 1990s to the late 2000s, which can be regarded as the "classical period" of renewable power policy. It was during this time that Spain emerged as a global leader in the field. Thanks to a generous support scheme known as the "special regime," which was developed and refined by governments on both sides of the political spectrum, Spain experienced rapid growth in first wind, then PV, and finally CSP installations, temporarily attaining some of the highest levels of capacity and electricity generation by these technologies in the world. Indeed, the amount of solar PV installed in particular greatly exceeded all plans and expectations, setting the stage for the next major phase in Spanish renewable power policy.

Chapter 4 examines the negative consequences of this first boom and the measures that were taken to address them, from the late 2000s to the mid-2010s. These years constitute the "dark ages" of renewable power policy in Spain. To close a growing gap between the receipts and outlays of the electric power system due in part to unexpectedly high support levels, successive governments, again of both major parties, first reduced the amount of remuneration that new renewable power plants could receive, then imposed a moratorium on new support, and finally scrapped the special regime altogether, replacing it with a complicated arrangement, the "specific remuneration regime," that threatened to reduce further the revenues of many existing producers. As a result, the deployment of new renewable installations slowed and then ground to a halt, and many existing plant owners were faced with the prospect of bankruptcy.

The late 2010s, the subject of Chapter 5, saw another reversal of fortune for the renewable industry, especially wind and PV, which enjoyed a virtual "renaissance." Under pressure to achieve EU renewable targets, the conservative government that had originally imposed the moratorium conducted a series of auctions that ultimately awarded potential support rights to nearly 9 gigawatts (GW) in new capacity, to be completed by 2020. Independently, a decline in costs combined with a desire for predictable electricity prices and more green power resulted in a steadily growing number of private decisions to build new power plants without any government support. As the decade came to an end, Spain experienced a second renewable power boom, led by a rapid doubling in the amount of PV.

Chapters 3 through 5 are primarily concerned with policies affecting utility-scale renewable power installations that sell all their output to the market. Chapter 6 examines the evolution of policy toward an alternative modality that has long been advocated in Spain: distributed renewable electricity generation that is largely consumed onsite by those who produce it, or what has popularly been known as self-consumption. In contrast to the policies examined so far, support for self-consumption has varied substantially across the major parties. Interest in self-consumption heightened in the early 2010s, as support for utility-scale renewables dried up. For much of the decade, however, investment was discouraged by the center-right governments of Mariano Rajoy, which developed and then adopted regulations that made self-consumption unprofitable. It was not until the late 2010s that the new Socialist government of Pedro Sánchez was able to lift those restrictions and investment began to take off.

Chapter 7 addresses the future prospects for renewable power in Spain. It begins by reviewing the reasons why Spain has adopted particularly ambitious renewable power targets for 2030. It then examines the broad challenges that Spain faces in achieving those targets and how the Sánchez governments have been attempting to overcome them. In particular, it examines whether sufficient generating capacity can actually be built, whether the desired capacity can be accommodated by the electric power grid, and whether all the power that the new capacity produces can be successfully integrated into the electricity market.

Chapter 8 provides an overarching political explanation of the patterns of policy outcomes observed in the preceding, primarily empirical chapters. To this end, it seeks to identify the key actors and structures in the policymaking process and the contributions that each has made. It confirms the importance of the core executive of the national government but finds that the aims sought by the party in power at any particular time cannot simply be derived from party ideology. Rather, a combination of external pressures (in the form of EU directives), largely objective and shared national imperatives (energy security, economic development, environmental sustainability, financial stability and solvency, etc.), and the efforts of interest groups must also be considered in order to understand the policies actually pursued. As a result, governments of the center-left and center-right have often adopted and followed similar policies, until international or domestic circumstances compelled a change of course.

## References

BP (British Petroleum). 2020. *Statistical Review of World Energy*. https://www.bp.com/en/global/corporate/energy-economics/statistical-review-of-world-energy.html.

Chari, Raj, and Paul M. Heywood. 2009. "Analysing the Policy Process in Democratic Spain." *West European Politics* 32, no. 1: 26–54.

CNMC (Comisión Nacional de los Mercados y la Competencia). 2021. "Información mensual de estadísticas sobre las ventas de régimen especial. Contiene información hasta diciembre de 2020" (23 Feb.). https://www.cnmc.es/estadistica/informacion-mensual-de-estadisticas-sobre-las-ventas-de-regimen-especial-contiene-52.

del Río, Pablo. 2016. "Implementation of Auctions for Renewable Energy Support in Spain." Report D7.1-ES (March). http://auresproject.eu/sites/aures.eu/files/media/documents/wp7_-_case_study_report_spain_1.pdf.

———. 2017. "Assessing the Design Elements in the Spanish Renewable Electricity Auction: An International Comparison." *Working Paper 6/2017*. Real Instituto Elcano (17 April). http://www.realinstitutoelcano.org/wps/portal/rielcano_en/contenido?WCM_GLOBAL_CONTEXT=/elcano/elcano_in/zonas_in/dt6-2017-delrio-design-spanish-renewable-electricity-auction.

del Río, Pablo, and Miguel Gaul. 2007. "An Integrated Assessment of the Feed-in Tariff System in Spain." *Energy Policy* 35, no. 2 (Feb.): 994–1012.

del Río, Pablo, and Pere Mir-Artigues. 2012. Support for Solar PV Deployment in Spain: Some Policy Lessons. *Renewable and Sustainable Energy Reviews* 16, no. 8 (Oct.): 5557–5566.

———. 2014. "A Cautionary Tale: Spain's Solar PV Investment Bubble, International Institute for Sustainable Development" (Feb.). https://www.iisd.org/gsi/sites/default/files/rens_ct_spain.pdf.

del Río González, Pablo. 2008. "Ten Years of Renewable Electricity Policies in Spain: An Analysis of Successive Feed-in Tariff Reforms." *Energy Policy* 36, no. 8 (Aug.): 2917–2929.

Heywood, Paul M. 1995. *The Government and Politics of Spain*. New York: St. Martin's.

———. 1999. "Power Diffusion or Concentration? In Search of the Spanish Policy Process." In *Politics and Policy in Democratic Spain: No Longer Different?*, edited by Paul M. Heywood, 103–123. London: Frank Cass.

Linares, Pedro, and Xavier Labandeira. 2013. "Renewable Electricity Support in Spain: A Natural Policy Experiment." *Economics for Energy WP 04/2013*. https://repositorio.comillas.edu/rest/bitstreams/18834/retrieve.

Medina, Iván, Joaquim M. Molins, and Luz Muñoz. 2016. "Interest Groups in Spain: Still Fragmented, Politicized, and Opaque?" Paper to be presented at the 2016 ECPR General Conference, Prague. https://ecpr.eu/Filestore/PaperProposal/da5ecbcf-82d7-4fb5-b2d9-681f9a2dd220.pdf.

Mir-Artigues, Pere. 2012. *Economía de la generación eléctrica solar: La regulación fotovoltaica y solar termoelétrica en España*. Civitas/Thomson Reuters.

———. 2013. "The Spanish Regulation of the Photovoltaic Demand-side Generation." *Energy Policy* 63 (Dec.): 664–273.

Mir-Artigues, Pere, Pablo del Río, and Emilio Cerdá. 2018. "The Impact of Regulation on Demand-side Generation: The Case of Spain." *Energy Policy* 121 (Oct.): 286–291.

Molins, Joaquim M., and Alex Casademunt. 1998. "Pressure groups and the articulation of interests." *West European Politics* 21, no. 4 (Dec.): 124–146.

Ritchie, Hannah, and Max Roser. 2020. "Renewable Energy." https://ourworldindata.org/renewable-energy.

Sáenz de Miera, Gonzalo, Pablo del Río, and Ignacio Vizcaino. 2008. "Analysing the Impact of Renewable Electricity Support Schemes on Power Prices: The Case of Wind Electricity in Spain. *Energy Policy* 36, no. 9 (Sept.): 3345–3359.

CHAPTER 2

# Laying the Groundwork: Early Efforts to Promote Renewable Power in Spain

## INTRODUCTION

Spain produced only a small amount of power from non-traditional sources prior to the turn of the millennium. But during the 1980s and 1990s, much of the groundwork was laid for the rapid buildup of wind and solar capacity that would take place in the first decade of the twenty-first century. The first efforts to promote renewable power were initiated in the early 1980s. Then, beginning in the middle of that decade, renewable power became for the first time a focus of energy planning. And in the early 1990s, Spain established the first comprehensive support mechanism, the so-called special regime, which would provide the basic framework for promoting renewable power until the early 2010s.[1]

## THE LEGACY OF FRANCO

Before examining the steps that Spain took to promote renewable power in the 1980s and 1990s, it is useful to review the renewable legacy of the Franco era. Electricity production and consumption in Spain grew dramatically after World War II as the country sought to modernize its economy. Between 1950 and 1980, total output rose from just 6,853 gigawatt-hours (GWh) to 110,483 GWh, and a significant share of that was provided by large hydroelectric power stations. To help meet the growing demand, installed hydroelectric capacity jumped from 1900 megawatts (MW) in 1950 to 13,700 MW in 1980, while the power they generated increased from 5000 GWh to 30,900 GWh over the same time period (Salmon 1995, 150 and 152).

By the early days of the new Spanish democracy, however, the potential for big hydroelectric plants had been largely exploited. In addition, hydroelectric output was unpredictable, varying potentially greatly from year to year depending on the amount of precipitation. As a result, an increasing share of Spain's electricity came from power plants fired by coal and fuel oil and, beginning in the late 1960s, from nuclear reactors. In contrast, the share provided by hydroelectric plants declined from a high of 84% in 1960 to just 28% in 1980. To meet the growing demand, Spain had to import increasing amounts of coal and oil (Salmon 1995, 150 and 152).

## The 1980s: Initial Efforts to Promote Renewable Power

Democratic Spain's efforts to promote renewable sources other than large-scale hydropower began in the early 1980s, not long after the new Constitution had been adopted. The timing was not random; these efforts formed part of Spain's response to the second oil shock and the resulting desire to enhance the security of the country's energy supplies by reducing dependence on energy imports in general and imported oil in particular. During this period, the government adopted the first legal framework for promoting renewable power and several sets of regulations for doing so. Then, during the second half of the decade, the government formally incorporated renewable power into its energy planning activities. Perhaps not surprisingly, given the relative immaturity of most renewable technologies, the results of these efforts were modest. Instead, Spain had to rely almost exclusively on coal and nuclear power in order to reduce its use of oil to generate electricity.

## Motivations

As the 1980s began, Spanish energy officials faced a daunting prospect. Electricity consumption had been growing by more than five percent per year on average, and it was expected to continue to rise at a comparable pace during the next decade. At the same time, the second oil shock underscored the risks of meeting that mushrooming demand with oil—27.5% in 1981—almost all of which was imported. Spain's foreign oil dependence, which accounted for 42% of the value of all imports that same year, contributed to a rough doubling of the country's trade deficit at the time of the oil shock. Even more importantly, it placed at risk the security of Spain's energy supply. Hence, the original center-right governments of the Union of the Democratic Center (*Unión de Centro Democrático* or UCD) deemed it imperative to reduce the country's oil consumption and import dependence at least in part by diversifying its energy sources (IEA 1981, 292; 1982, 302–303; Lancaster 1989, 48; Ley 82/1980; BOCG 1984). At this point, interestingly, environmental concerns played little or no role in shaping renewable power policy. As a 1981 International Energy Agency (IEA) report noted, "the impact of coal (as a substitute for oil) on the environment... does not create major difficulties" (IEA 1981, 292; see also del Río González 2008, 2918).

## Initial Actions Taken

Part of Spain's response to the second oil shock was to maximize the utilization of its hydroelectric potential. In 1982, the UCD government adopted a regulation promoting the construction of large-scale (greater than 5 MW) hydroelectric power plants, and it established plans to increase that capacity by some 5500 MW over the next decade (Real Decreto 1544/1982; IEA 1982, 304).

For the first time, in the 1980 Law on Energy Conservation, the government singled out for support small-scale hydropower (up to 5 MW) and what was called self-generation (*autogeneración*) of electricity by producers who directly consumed at least some of the power they produced. The law established the right of self-generators and owners of small-scale hydroelectric plants to connect to the grid and to transfer any excess electricity to the local distribution company at a regulated price. The law also provided for a variety of other financial incentives, such as subsidized loans and tax deductions. It should be noted, however, that autogenerators were not limited to using renewable sources; they included co-generation with fossil fuels and other sources of heat as well (Ley 82/1980).

### First Planning for Renewable Power

Over the following years, successive governments of different parties—the Socialists under Felipe González won the general election of 1982 and took office that December—issued a series of detailed regulations (*Real Decreto* or RD) and ministerial orders that were needed to flesh out the broad framework provided by the Energy Conservation Law. Some of the more important details that remained to be determined were the administrative and technical rules for connecting to the grid, how and by whom the sales price would be set, and the precise nature of the incentives and supports to be provided (Real Decreto 1217/1981; Real Decreto 907/1982; Orden de 28 de julio de 1982; Orden de 5 de septiembre de 1985).

In the meantime, the promotion of renewable energy, including renewable power, became central to the country's energy planning process for the first time. The new Socialist government undertook a comprehensive review of the existing National Energy Plan (*Plan Energético Nacional* or PEN), issuing a new plan in 1983 that was formally adopted in June the following year. The PEN 1983 in turn called for the elaboration of a more focused Renewable Energies Plan (*Plan de Energías Renovables* or PER), which would be the first of its kind in Spain (BOCG 1984, 452).

The PER was issued in 1986 and covered the period through 1992. It established concrete goals, resources, and programs of action. The overall goal was to increase the contribution of renewable energy, not including large-scale hydroelectric power, to 0.91% of primary energy consumption in 1988 and then to 2.73% by 1992, and the plan set more specific targets for individual technologies. To achieve this goal, Spain would have to invest 55 billion pesetas (about $400 million in 1986 dollars) between 1986 and 1988 alone. In addition, new energy resources would receive the largest share, about one-quarter, of the government's energy research and development budget—more than conservation, nuclear power, or coal alone—for a total of an additional 6.9 billion pesetas (about $50 million) for the period 1985 through 1987 (González 2013; IEA 1986, 350, 367, 361; Jacobs 2012, 53).[2]

Because the targets for the two most cost-effective technologies at the time, biomass and small-scale hydroelectric power, were reached in 1988, a second PER was adopted in 1989 for the period through 1995. The new plan raised the overall renewable target to 4%, which would require a total investment of 145 billion pesetas (about $1.2 billion in 1989 dollars), including 35 billion pesetas (about $290 million) in public aid (IEA 1989, 545; IEA 1990, 401; Jacobs 2012, 53).

*Impact*

The results of these efforts were mixed. By 1990, renewable energy, not including large-scale hydropower, had already reached 2.578 million tonnes of oil equivalent (MTOE), or approximately 2.9% of primary energy consumption. Of that total, however, more than 90% was provided by biomass. Renewable power, again not including large-scale hydropower, accounted for just one percent of all electricity production, at 1559 gigawatt-hours (GWh), and just five percent of all renewable energy. Nearly 90% of that was provided by small-scale hydroelectric installations, which had grown to 458 MW of capacity, out of an overall renewable power capacity of 495.2 MW. In contrast, total wind and solar photovoltaic (PV) power potential had reached just 7.2 MW and 3.16 MW, respectively, reflecting the relative immaturity and costliness of those technologies (BOCG 1991, 179; BP 2020).

Instead, Spain had relied primarily on nuclear power and coal to substitute for oil in the generation of electricity during the 1980s. The country's nuclear capacity tripled with the opening of five more reactors. And coal consumption increased by two-thirds, from 12.4 MTOE in 1980 to 19.2 MTOE in 1990, with much of the addition being used to generate electricity (BP 2020; IEA 1981, 292).

Nevertheless, the foundation had been laid for subsequent growth in renewable power capacity. Indeed, Spain was poised to become the home of the two largest wind farms in Europe—one of 20 MW and one of 10 MW—which entered service in 1992 (Salmon 1995, 161). All that was needed to realize this growth was the right set of incentives.

## The 1990s: The First Systematic Support for Renewable Power

Those incentives would soon be provided with the adoption of even more ambitious goals and the first comprehensive scheme for supporting renewable power in the first half of the 1990s. By then, the original motives for promoting renewable energy—reducing foreign oil dependence, increasing self-sufficiency, and diversifying energy supplies—had been joined by a new set of concerns: the environmental impact of the energy sector. As a 1991 planning document noted, "Production and consumption of energy are one of the principal causes of air pollution, and recently, $CO_2$ emissions have acquired a special relevance for its contribution to the greenhouse effect... the objective of limiting $CO_2$ emissions is probably one of the most important elements for justifying our new plans." In addition, government officials were beginning to appreciate the potential commercial opportunities represented by investment in renewable power (BOCG 1991, 154, 174–175).

### *The 1991 National Energy Plan*

In 1991, the Socialist government adopted a new National Energy Plan for the period through the end of the decade, *Plan Energético Nacional 1991–2000*. The new plan in turn contained a Renewable Energy Program that sought to build on the 1989 plan. For renewable power, again not including large-scale hydropower, the program projected a 240% increase in renewable capacity, from 495.4 to 1684 MW, and a 270% increase in renewable electricity production, from 1559 to 5738 GWh, by 2020, and these goals were thought to be conservative. If achieved, renewables would cover 3.2% of Spain's electricity demand, up from 1.2% in 1990 (BOCG 1991, 160, 177, 179).[3]

The projected growth in capacity and power generation, however, was unevenly distributed across alternative technologies, reflecting major differences in cost-competitiveness. The greatest potential was still thought to lie with small-scale hydroelectric power (now defined as 10 MW or less), whose capacity would rise from 458 to 1237 MW, representing nearly two-thirds of all capacity growth in renewables. Indeed, virtually all the increase in hydropower would take the form of small-scale projects, reflecting the realization that virtually all large-scale hydropower potential had already been exploited. Next on the list was the production of electricity from waste, the capacity and power generation of which would increase ten-fold, to 266 MW and 1453.5 GWh, respectively. Only then came wind power, although the plan did call for large increases in capacity and output in relative, if not absolute, terms, to 175.2 MW and 421 GWh. And the capacity and output of solar photovoltaic electricity were not even expected to double from their already miniscule levels (BOCG 1991, 53, 179).

The required investment in renewable energy, totaling 334 billion pesetas (about $3.2 billion in 1991 dollars), was expected to be similarly skewed. 46.7% (156 billion pesetas), would go to minihydro. Another 23.9% (80 billion pesetas) would be invested in applications of waste, although some of that would be for non-electrical uses. In contrast, wind would receive just 8.1% (27 billion pesetas) and solar PV only 1.8% (6 billion pesetas). Nevertheless, the amount that would come in the form of public support was more evenly distributed, with 14.7 billion pesetas going to minihydro, 20.3 to waste, 11.8 to wind, and 2.95 to PV (BOCG 1991, 192–193).

### *The Original Special Regime*

Within a few years, however, it became clear that more support would be needed if the targets in the 1991 energy plan were to be achieved. In particular, the incentives created in the 1980s, mainly to promote minihydro, had proven inadequate for stimulating the growth of co-generation and renewable power to the desired extent. As a result, the Socialist government developed a new regulation, *Real Decreto* 2366/1994, that, for the first time, provided a comprehensive support scheme for renewable sources of electricity involving what was effectively a feed-in tariff. Adopted in December 1994, the regulation established a "special regime" for power plants of up to 100 MW based on co-generation and renewable sources and hydroelectric plants of up to 10 MW. Qualifying installations would receive access to the grid and the right to sell any surplus production to power distribution companies under contracts lasting at least five years. The price would be determined by a complex formula based primarily on the avoided costs of conventional power generation, while also taking into account factors such as time of generation and environmental and social benefits, but otherwise did not differentiate between alternative non-hydro renewable technologies. The distribution companies would pass these costs along to their customers (Real Decreto 2366/1994; Jacobs 2012, 69–70, 104, 179).[4]

Overall, the regulation explicitly sought to strike a balance between ensuring an adequate profit for producers and avoiding significant increases in the electricity tariffs paid by consumers. The government initially set the feed-in tariffs at less than double the average cost of electricity and gradually reduced them thereafter. Given that renewable power was expected to reach just 3.2% of all output, its impact on electricity bills would likely be negligible (Jacobs 2012, 179; Eikeland and Saeverud 2007, 27; OECD 1999, 30).[5]

*Outcomes*

The new feed-in tariff contributed to a surprisingly rapid growth in renewable power at a modest cost. Indeed, many of the targets for 2000 set in the 1991 energy plan were exceeded by 1998. Overall, at that time, renewable capacity had reached 2635 megawatts (versus the goal of 1684 MW) and renewable output stood at 8784 GWh (versus 5738 GWh), which represented 4.5% of all electricity produced (versus 3.2%), not including the contribution of large-scale hydropower. As expected, the greatest amount of growth occurred in small-scale hydroelectric plants, which reached 1510 GW (versus 1237 GW) of capacity and 5607 GWh (versus 3853 GWh) of output (IDAE 1999, 43).

What was not expected was how quickly wind and biomass would increase. Wind reached 834 MW of capacity in 1998, nearly five times the goal of just 175 MW, and 1437 GWh in output (versus only 421 GWh) and would continue to grow—by a remarkable 750 MW in 1999 alone (IEA 2001, 107). Biomass, which had not even been given a separate target for electricity production in PEN 1991, grew to 189 GW of capacity and an impressive 1139 GWh of output, reflecting its high capacity factor. Only electricity production from solid waste fell short of its targets, at 94 MW (versus 265 MW) and 586 GWh (versus 1454 GWh). Solar photovoltaic remained far behind, though it too exceeded its admittedly modest targets, reaching 8.7 MW of capacity and 15.3 GWh of output (IDAE 1999, 43).[6]

## Conclusion

As made clear in this chapter, the story of determined efforts by Spanish governments to promote renewable power begins in the early 1980s, with the adoption of the first legal framework and supporting regulations. That was followed by the formal incorporation of renewable power into Spain's energy planning process and then, in the mid-1990s, by the adoption of the first comprehensive support scheme. As a result, although renewable power still represented only a small portion of Spain's electricity generation, it was growing rapidly in relative terms as the decade came to a close. Indeed, the growth in wind power was accelerating, roughly doubling every year in the mid- to late 1990s. The stage was finally set for non-traditional renewable sources to provide a major share of Spain's electricity supply if provided with the right incentives.

During this period, the motives for promoting renewable power also steadily expanded. Initially, doing so was justified by heightened concerns about the security of energy supplies and the costs of energy imports in the context of a high degree of dependence on foreign energy sources. Subsequently, environmental concerns, including the impact of greenhouse gas emissions from the burning of fossil fuels, assumed equal prominence. And as the renewable sector began to grow, political leaders also began to consider the potential economic benefits.

Finally, this period provides early examples of the cross-party consensus that was to characterize renewable power policy during later years. The Socialist governments under González steadily built upon the initial steps taken by the conservative governments that preceded them. Later, the successor center-right People's Party government of José Maria Aznar, elected in 1996, continued the policies it inherited and, as we shall see in the next chapter, then built upon them to create the promotional scheme that would yield Spain's initial boom in wind and solar power.

## Notes

1. The focus of this chapter and of the book as a whole is on non-traditional sources of renewable electricity, such as wind, solar, biomass, and small-scale hydroelectric plants. Large-scale hydroelectric power has also been an important source of electricity in Spain, but its potential was almost fully exploited by the time the new Constitution was adopted in 1978 and thus its contribution to overall electricity generation has been relatively constant, subject mainly to variation in rainfall.

2. 1 euro equaled 166.386 pesetas at the time of replacement (1999). Exchange rates for earlier years are derived from https://www.pounds terlinglive.com/bank-of-england-spot/historical-spot-exchange-rates/usd/USD-to-ESP.
2. The plan also called for increasing the share of electricity from self-generation from 4.5 to 10%, but the majority of that would come from co-generation using conventional sources of energy.
3. The new regulation formally replaced the three regulations adopted in the early 1980s to promote hydroelectric power and self-generation: *Real Decreto* 1217/1981, *Real Decreto* 907/1982, and *Real Decreto* 1544/1982. Installations larger than 100 MW (or 10 MW in the case of hydroelectric plants) would continue to be regulated under a separate framework that determined the electricity tariff, *Real Decreto* 1538/1987. The new regulation also created a registry in which all participating installations would have to be inscribed.
4. In the 1990s, Spain also maintained assistance programs for small-scale renewable power projects that faced particularly challenging circumstances (Varela and González 1999, 36–37; Real Decreto 615/1998).
5. The term capacity factor refers to the ratio of the actual electrical energy output of a given installation over a given period of time to the maximum possible electrical energy output over that period.

## References

BOCG (Bulletin Oficial de las Cortes Generales). 1984. *Plan Energético Nacional 1983* (14 May). http://www.congreso.es/public_oficiales/L2/CONG/BOCG/E/E_042.PDF.

———. 1991. *Plan Energético Nacional 1991–2000* (13 Sept.). http://www.congreso.es/public_oficiales/L4/CONG/BOCG/E/E_169.PDF.

BP (British Petroleum). 2020. *Statistical Review of World Energy*. https://www.bp.com/en/global/corporate/energy-economics/statistical-review-of-world-energy.html.

del Río González, Pablo. 2008. "Ten Years of Renewable Electricity Policies in Spain: An Analysis of Successive Feed-in Tariff Reforms." *Energy Policy* 36, no. 8 (Aug.): 2917–2929.

Eikeland, Per Ove, and Ingvild Andreassen Saeverud. 2007. "Market Diffusion of New Renewable Energy in Europe: Explaining Front-Runner and Laggard Positions." *Energy & Environment* 18, no. 1: 13–36.

González Vázquez, José Manuel. 2013. "Fomento de las energias renovables en Espana" (21 Oct.). https://www.eoi.es/blogs/josemanuelgonzalezvazquez/2013/10/21/fomento-de-las-energias-renovables-en-espana-lecciones-aprendidas-futuro/.

IDAE (Instituto para la Diversificación y Ahorro de la Energía). 1999. *Plan de Fomento de las Energías Renovables en España* (Dec.). https://www.idae.es/uploads/documentos/documentos_4044_PFER2000-10_1999_1cd4b316.pdf.
IEA (International Energy Agency). 1981. *Energy Policies and Programmes of IEA Countries: 1981 Review*. Paris: OECD.
———. 1982. *Energy Policies and Programmes of IEA Countries: 1982 Review*. Paris: OECD.
———. 1986. *Energy Policies and Programmes of IEA Countries: 1986 Review*. Paris: OECD.
———. 1989. *Energy Policies and Programmes of IEA Countries: 1989 Review*. Paris: OECD.
———. 1990. *Energy Policies and Programmes of IEA Countries: 1990 Review*. Paris: OECD.
———. 2001. *Energy Policies of IEA Countries: Spain 2001 Review*. Paris: OECD.
Jacobs, David. 2012. *Renewable Energy Policy Convergence in the EU: The Evolution of Feed-in Tariffs in Germany, Spain, and France*. London and New York: Routledge.
Lancaster, Thomas D. 1989. *Policy Stability and Democratic Change: Energy in Spain's Transition*. University Park and London: The Pennsylvania State University Press.
Ley 82/1980, de 30 de diciembre, sobre conservación de energía. 1980. https://www.boe.es/eli/es/l/1980/12/30/82.
OECD (Organization for Economic Cooperation and Development). 1999. *Spain: Regulatory Reform in the Electricity Industry*. http://www.oecd.org/regreform/sectors/2497385.pdf.
Orden de 28 de julio de 1982 por la que se desarrolla el Real Decreto 1217/1981, de 10 de abril, para el fomento de la producción hidroeléctrica en pequeñas centrales. 1982. https://www.boe.es/eli/es/o/1982/07/28/(2).
Orden de 5 de septiembre de 1985 por la que se establecen normas administrativas y técnicas para funcionamiento y conexión a las redes eléctricas de centrales hidroeléctricas de hasta 5.000 KVA y centrales de autogeneración eléctrica. 1985. https://www.boe.es/eli/es/o/1985/09/05/(1).
Real Decreto 1217/1981, de 10 de abril, para el fomento de la producción hidroeléctrica en pequeñas centrales. 1981. https://www.boe.es/eli/es/rd/1981/04/10/1217.
Real Decreto 907/1982, de 2 de abril, sobre fomento de la autogeneración de energía eléctrica. 1982. https://www.boe.es/eli/es/rd/1982/04/02/907.

Real Decreto 1544/1982, de 25 de junio, sobre fomento de construcción de centrales hidroeléctricas. 1982. https://www.boe.es/eli/es/rd/1982/06/25/1544.

Real Decreto 2366/1994, de 9 de diciembre, sobre producción de energía eléctrica por instalaciones hidráulicas, de cogeneración y otras abastecidas por recursos o fuentes de energía renovables. 1994. https://www.boe.es/eli/es/rd/1994/12/09/2366.

Real Decreto 615/1998, de 17 de abril, por el que se establece un régimen de ayudas y se regula su sistema de gestión en el marco del Plan de Ahorro y Eficiencia Energética. 1998. https://www.boe.es/eli/es/rd/1998/04/17/615.

Salmon, Keith. 1995. *The Modern Spanish Economy.* 2nd ed. London and New York: Pinter.

Varela, M., and R. González. 1999. "The Spanish Wind Situation and Comparison with Portugal and The Netherlands." *Informes Tecnicos Ciemat* 884. Instituto de Estudios de la Energia (March). https://inis.iaea.org/collection/NCLCollectionStore/_Public/38/106/38106941.pdf.

CHAPTER 3

# The Classical Period: Renewable Power Takes Off

## INTRODUCTION

During the decade from the late 1990s to the late 2000s, successive Spanish governments developed and refined a generous system for promoting investment in renewable power. This support system, the "special regime," was based primarily on a combination of regulated tariffs and premiums on top of the market price of electricity that facility owners could choose between. It was designed to ensure that investors would receive a reasonable return on their investments while trying to minimize the financial burden on electricity users, who would ultimately pay the costs of the support system. I term this the "Classical Period" in the politics of renewable power in Spain, as it saw the first significant buildup in renewable generating capacity and electricity output.

Although political backing for the special regime remained steady, despite a change in the party in power following the 2004 general election, one can nevertheless discern three main phases in the evolution of the support system during this time frame. The first phase dates to 1997 and 1998, with the introduction of a new Electricity Sector Law and the promulgation of the first regulations (*Real Decreto*) for the new special regime by the People's Party government headed by José María Aznar. Then, shortly before the 2004 election, the second Aznar government issued a new set of regulations that attempted to address problems with the original ones as well as rising renewable targets. Finally, in 2007, the Socialist government of José Luis Rodríguez Zapatero took its own turn at fine-tuning the regulations governing the special regime in order to respond to persisting criticisms and further increases in the anticipated need for renewable power.

This chapter seeks to answer three principal questions. Why did successive Spanish governments seek to promote renewable power during this period? What measures did they adopt and for what reasons? And what were the results of their efforts?

## First Steps: The New "Special Regime"

In 1996, the Socialists, who had done so much to promote renewable energy, were voted out of office after 14 years in power. One of the top priorities of the new center-right government was to introduce market-based reforms in the electricity sector, as mandated by the European Union. Nevertheless, the level of support for renewable power did not flag, as the Aznar government updated the original special regime established by the Socialists in 1994 to be compatible with a liberalized electricity market. Indeed, the level of support for new renewable power installations increased in some ways, reflecting growing pressure both at home and abroad to increase the share of renewables in Spain's energy mix. The new special regime was especially successful at maintaining, and even accelerating, the rapid growth in wind power that had occurred in the previous several years.

### *Motives for a New Push to Promote Renewable Power*

Even as the feed-in tariff adopted in 1994 was yielding positive results, pressure was mounting to do even more to promote renewable energy in Spain. One source of pressure was continued concern about Spain's dependence on foreign energy sources. Imports still provided for more than 70% of Spain's energy needs, despite the growth in nuclear power and renewable sources. Indeed, energy imports grew by more than 60% between 1985 and 1998. In the meantime, environmental concerns, especially the need to reduce greenhouse gas emissions, had grown even stronger. In 1998, Spain signed the Kyoto Protocol, which would initially require it to limit the growth of its emissions to no more than 15% above the 1990 levels, and renewable energy was regarded as critical for meeting this target (IDAE 1999, 1, 7–14).

These pre-existing motives were reinforced by a growing realization of the potential positive socioeconomic impacts of renewable energy. Investment in renewables was increasingly seen as useful for promoting technological advancement and industrial competitiveness, generating businesses and employment, and spurring regional development, since many of the new jobs would be in areas with relatively few employment opportunities. By one 1999 government estimate, investment in renewables could create some 45,000 new positions by 2010 and a total of 84,000 by 2020 (IDAE 1999, 16–25).

At the same time, Spain was under pressure from the European Union (EU) to do more in the area of renewables. In 1996, the European Commission proposed doubling the share of renewables in gross energy consumption from the then current level of 6% to 12% in 2010, a target that was endorsed by the European Council the following year (CEC 1996; EC 1997).

Meanwhile, estimates of Spain's renewable power potential only continued to grow. According to a 1999 government estimate, the potential for wind power stood at 15 gigawatts (GW), and the amount of solar photovoltaic (PV) capacity that Spain could build was limited only by cost. Other renewable sources could also contribute substantial, if lesser, amounts of generating capacity, including small-scale (mini) hydroelectric (2400 megawatts (MW)) and mid-size hydroelectric plants of between 10 and 50 MW (10,400 MW) (IDAE 1999, 56–59).

Whatever Spain did to promote renewable power, however, would have to be compatible with the liberalization of the electricity sector upon which the country was about to embark. In 1996, the EU adopted common rules (Directive 96/92/EC) for the internal electricity market, which called for opening up for competition those areas—production, supply, and sales—that were not subject to natural monopoly. The next year, the Aznar government transposed these rules into the Spanish legal code with the passage of the Electricity Sector Law, which provided for the gradual liberalization of the retail electricity market, competition in generation, the legal separation of network business from those engaged in generation and supply, independent regulators, and non-discriminatory access by generators to the network (Ley 54/1997; for a more detailed discussion, see OECD 1999, 19–23).

Nevertheless, the EU directive permitted giving priority to the production of electricity from renewable sources for environmental reasons. Likewise, the Spanish law sought to make liberalization compatible with the achievement of other goals, such as increasing energy efficiency, reducing consumption, and protecting the environment. These provisions left considerable scope for special measures to promote renewable power.

### Efforts to Promote Renewable Power

The Aznar government's initial efforts to do so therefore involved two steps. First, it provided the basic framework for a support system that would be compatible with a liberalized market in the 1997 Electricity Sector Law. Then, it spelled out specific rules and procedures in a detailed set of regulations.

The new law drew a distinction between an ordinary regime and an updated version of the special regime already in place. The former governed the free market in power generation by traditional sources of electricity, such as coal, nuclear, and large hydroelectric. The latter applied to installations of up to 50 MW in capacity (10 MW for hydroelectric plants)—down from the 100 MW limit contained in the 1994 special regime—that used renewable energy, non-renewable waste, and co-generation in the case of self-producers. Renewable producers that qualified for the special regime had the right to connect to the grid and to sell the electricity they generated—the excess electricity in the case of self-producers—to the system in return for a premium, on top of the market price, set by the government. The premium was justified as a cost of securing and diversifying Spain's electricity supply and would take into account the environmental contributions as well as the need for investors to achieve a reasonable rate of return. The law also provided for the possibility of a higher premium for solar power as well as premiums for renewable installations of more than 50 MW (between 10 and 50 MW in the case of hydroelectric plants). Finally, the law called for the preparation of a plan for the promotion of renewable energy that the government would take into account when setting the premiums, with the aim of covering at least 12% of total energy demand with renewable sources by 2010, in line with the new EU target (Ley 54/1997).

The passage of the law in late 1997 was followed by the promulgation of the regulations, *Real Decreto* 2818/1998, needed to adopt the special regime to the introduction of the free market in generation, beginning

January 1 of 1999. The goal was to promote renewable and other special regime production facilities without creating discriminatory conditions that could limit competition. The regulations reconfirmed the right of renewable generators to connect to the distribution grid and transfer their production to distributors in return for the market price plus an incentive, under contracts of at least five years. Distributors would pass the additional cost on to the recently established National Energy Commission (*Comisión Nacional de la Energía* or CNE), which would then include them in the tariffs paid by consumers (del Río González 2008, 2918).

The most novel feature of the regulations was the presence of two alternative forms of support. Producers could choose between receiving a premium on top of the market price of electricity or a fixed total price for each kilowatt-hour sold.[1] This "double option" was intended to encourage gradual participation in the electricity market while mitigating the risk for renewable generators. In addition, there were no limits to how long a producer could receive support, although this provision was confined to the preamble and the incentives could be adjusted every year by the government and thoroughly revised every four years in an attempt to maintain a balance between stimulating investment and minimizing the burden on consumers. Producers were also responsible for the cost of connecting to and, if necessary, reinforcing the grid (Real Decreto 2818/1998; del Río González 2008, 2924; Mir-Artigues 2012, 344–347).

Finally, the regulations differentiated support levels by technology and installation size. They offered particularly generous premiums and fixed total prices to solar producers, especially installations of no more than 5 kilowatts (kW), and a small premium was provided for renewable plants greater than 50 MW.

As the new support system went into effect, the Aznar government completed the promotion plan called for in the 1997 law. Prepared by the Institute for Energy Savings and Diversification (*Instituto para la Diversificación y Ahorro de la Energía* or IDAE) working under the Ministry of Industry and Energy and approved at the very end of 1999, the plan established targets for 2010 and identified measures and strategies for attaining them. If all the targets were achieved, renewable power output would nearly double, from 39.5 terawatt-hours (TWh) in 1998 to 76.6 TWh in 2010, when it would represent 29.4% of total electricity generation, with most of the growth coming in non-hydroelectric sources. $CO_2$ emissions in the power sector in 2010 would be reduced between 14.56

and 36.55 million tonnes, depending on whether coal or natural gas was displaced by renewables. Overall, renewable sources would provide for 12.3% of Spain's primary energy consumption, slightly exceeding the goal established in the law (IDAE 1999).

In the electricity sector, the plan placed particular emphasis on wind and biomass. The generating capacity of both would increase 10-fold, and together they would account for nearly 90% of the growth in power output. Minihydro (less than 10 MW) and solid waste would each contribute another three to four percent of the increase, while solar and biogas would make only marginal contributions. For the first time, however, the plan provided separate targets for photovoltaic (PV) and concentrated solar power (CSP), the two main solar technologies (Tables 3.1 and 3.2).

Table 3.1 Renewable power capacity and targets, 1998–2010 (MW)

|  | 1998 Actual | Targets for 2010 | | 2010 Actual |
|---|---|---|---|---|
|  |  | 1999 | 2005 |  |
| Minihydro (<10 MW) | 1510 | 2230 | 2199 | 2026 |
| Wind | 834 | 8974 | 20,155 | 19,700 |
| Biomass | 189 | 1897 | 2039 | 709 |
| Solar PV | 8.7 | 144 | 400 | 3830 |
| CSP | 0 | 200 | 500 | 532 |

*Sources* IDAE (1999, p. 61), IDAE (2005, p. 327), and CNMC (2021, Table 1.2)

Table 3.2 Renewable power generation and targets, 1998–2010 (GWh)

|  | 1998 Actual | Targets for 2010 | | 2010 Actual |
|---|---|---|---|---|
|  |  | 1999 | 2005 |  |
| Minihydro (<10 MW) | 5607 | 6912 | 6692 | 6743 |
| Wind | 1437 | 21,538 | 45,511 | 43,142 |
| Biomass | 1139 | 13,949 | 14,015 | 3140 |
| Solar PV | 3.8 | 176 | 609 | 6402 |
| CSP | 0 | 459 | 1298 | 621 |

*Sources* IDAE (1999, p. 37), IDAE (2005, p. 327), and CNMC (2021, Table 1.2)

What would it take to achieve these targets? The plan identified a number of means for promoting renewables, including public investment in research and development, financing infrastructure, harmonizing regulations at the regional and local levels, especially those concerning environmental impacts, and simplifying administrative procedures, such as grid connections for small plants. But a substantial input would be financial.

The plan estimated that required investment in the renewable power sector would total 1525 billion pesetas (9.16 billion euros) between 1999 and 2006, of which only 69.4 billion pesetas (417 million euros) would take the form of public subsidies. Other subsidies and fiscal incentives would add up to another 167 billion pesetas (1.00 billion euros). A majority of the subsides, including those for accompanying measures, would be covered by the European Union, with the remainder divided among the national, regional, and local governments of Spain.[2]

Instead, the greatest amount of support—434 billion pesetas (2.61 billion euros)—would come in the form of the premiums and fixed prices established in the new law and regulations for the special regime. By 2006, payments would total 123 billion pesetas (739 million euros) per year. Given the anticipated contributions of wind and biomass, those technologies were expected to receive the lion's share of the support (59% and 23%, respectively), while biomass would account for more than half of the subsidies and fiscal incentives.

### Impact of the New Special Regime

The implementation of *Real Decreto* 2818/1998 was followed by an acceleration in the deployment of wind power. The total amount of installed wind capacity approximately doubled in 1999 and then grew at an average rate of more than 1000 MW annually for the next four years, reaching more than 6000 MW in 2003. By then, Spain had more wind power capacity than any country but Germany. In contrast, biomass and minihdyro increased less rapidly than anticipated, raising questions about whether the targets enshrined in the plan could be met (IEA 2005, 114). And as expected, the amount of solar power remained small, although PV was beginning to take off, nearly doubling in capacity three years in a row, thanks to the generous support levels established in 1998 and falling costs. At the end of 2003, total grid-connected PV capacity stood at more than 10 MW, though not a single CSP plant had been built.

Despite the growth, the financial burden imposed on consumers by the special regime remained modest. In 2003, the total value of the support paid reached 1200 million euros. Subtracting the average cost of electricity, moreover, yielded a net cost of just over 620 million euros. Thus the effective additional cost of electricity for consumers was 0.26 eurocents per kilowatt-hour (kWh) or less than nine percent of the final electricity price (del Río and Gaul 2007, 1006).

Meanwhile, the investment in renewables was already regarded as having substantial socioeconomic benefits. The total number of companies operating in the sector jumped from 500 in the late 1990s to around 1400 in 2004, of which some 500 were associated with wind power. The IDAE estimated that those companies in turn had created tens of thousands of new positions, both direct and indirect: over 96,000 in wind, 2600 in minihydro, 3200 in CSP, and 2400 in PV. And these jobs were located to a large extent in rural areas where there had been limited opportunities for employment (IDAE 2005, 25–27). More generally, del Río and Gaul found, the support provided by the new special regime led to the development of a renewable power "technoinstitutional complex" made up of learning networks between producers, equipment suppliers, local communities, policymakers, and non-governmental organizations, whose mutual interactions created a positive feedback loop, especially in the case of wind (del Río and Gaul 2007, 1009).

## INITIAL ADJUSTMENTS UNDER THE AZNAR GOVERNMENT

*Real Decreto* 2818/1998 governed the special regime for more than five years. It soon became clear, however, that some adjustments would be needed, and in 2004, the Aznar government issued an entirely new set of regulations shortly before its unexpected defeat in the general elections that March. *Real Decreto* 436/2004 sought to provide greater investment certainty while doing more to encourage generators to sell directly to the wholesale market and limiting the potential burden to consumers. To do so, it formally linked support levels to electricity prices, provided an additional market incentive, and placed new limits on the amount of support that could be provided, although initial support levels were generally higher than under the previous regulations.

## Motives for Making Adjustments

One motivation for the development of new regulations was the intrinsic limitations of the existing regime. Some observers argued that the support levels were too low in some cases, and indeed, those for biomass were increased substantially (19.4% for premiums and 11.0% for fixed prices) during what turned out to be the last full year of operation of *Real Decreto* 2818/1998. Of more general concern were the annual updating of support levels for existing as well as new plants and the fact that contracts with distributors need run no longer than five years. It was argued that these provisions created uncertainty and risks for investors, and resulted in lending institutions demanding higher levels of interest (del Río and Gaul 2007, 999, 1010–1011). In fact, apart from a decrease in the second year—around 9% for premiums and 5.5% for fixed prices— the support levels remained virtually constant prior to the general review of 2003, and at that time, only premiums for wind, which was having no trouble attracting investment, saw a significant decline (8.0%). A related source of concern was that few producers were taking advantage of the market option, precisely because choosing the fixed price provided greater investment security and, as it turned out, higher profitability (Jacobs 2012, 202, 145; Eikeland and Saeverud 2007, 28). Thus the goal of eventually transitioning to a truly competitive power market was impeded.

Finally, there were a number of other regulatory issues that either *Real Decreto* 2818/1998 did not resolve or that emerged as a result of its success. For example, as the share of intermittent renewables, and especially wind power, in the electricity mix grew, concerns arose about the impact on the stability of the grid and power supplies. There were no limits on what utilities could charge for connecting producers to the grid, an issue that required the adoption of a separate regulation (Real Decreto 1663/2000). It was unclear who was responsible for increasing grid capacity in order to accommodate new power plants. And rules were needed to regulate self-producers (del Río González 2008, 2918, and 2924; del Río and Gaul 2007, 1000–1001; del Río and Unruh 2007, 1507; del Río and Mir-Artigues 2012, 559).

Meanwhile, Spain's objectives for renewable power continued to grow. In 2001, the EU set an indicative target for Spain of 29.4% of electricity from renewable sources by 2010, up from just 20% (mostly hydroelectric power) in 1997 (Directive 2001/77/CE). The next year, this target—rounded to 30%—was incorporated into Spain's electricity and gas infrastructure plan for 2002–2011. Because domestic electricity consumption was growing faster than expected, however, the forecast for installed renewable generating capacity was much greater than anticipated by the plan just three years before. As before, the bulk of the new capacity would be provided by wind and biomass, which saw their forecasts increase by 45% (13,000 MW) and 63% (3100 MW), respectively. The targets for solar PV, CSP, solid waste, and biogas stayed the same, and together accounted for just 3.6% of the forecast total renewable capacity, even when excluding large hydropower (MINECON 2002).

Finally, national politics may have played a role in determining the timing, if not the content, of the new regulations. The People's Party seemed anxious to approve them before the general election scheduled for March 2004, which they did by two days. As a result, voices in the renewable sector noted the haste with which the regulations were developed and the lack of transparency in the process (APPA 2004a, 2004b).

### *Taking Action:* Real Decreto 436/2004

The new regulations, contained in *Real Decreto* 436/2004, were much more detailed and extensive than those they replaced, covering some 21 pages in the official bulletin of the government. They continued to offer two basic options to renewable generators: selling their production (1) directly to the wholesale market in return for the market price and a premium or (2) to a distributor for what was now called a regulated tariff. They also established a slightly higher set of technology-specific targets based on the 2002 electricity infrastructure plan (Real Decreto 436/2004).

Many of the details sought to address the concerns that had been raised. First, they dealt with investor risk and uncertainty by linking support levels to the average price of electricity (*tarifa eléctrica media* or TEM), updated annually. The government would be able to exercise discretion only when it came to revising the support levels every four years, starting in 2006, and any changes would apply only to new installations, not retroactively to existing ones.

At the same time, several provisions sought to limit the potential burden on consumers. Although support continued to be guaranteed for the entire lifetime of a power plant, a one-time reduction in the regulated tariff would occur after a certain number of years, depending on the technology and capacity. Initial support periods ranged from as many as 25 years (solar PV, CSP, and small hydro) to as few as five (hydro plants from 10 to 25 MW). In addition, once the new capacity target for a particular technology was achieved, the level of support for new plants would be reviewed. No attempt was made, however, to tie support levels to likely cost reductions as a result of learning and economies of scale.

In order to encourage more extensive use of the market option, the regulations included a new market participation incentive set at 10% of the TEM. In addition, although most support levels for the initial period of operation went up, the total compensation (premium plus incentive) for those selling to the market tended to increase more in both absolute and relative terms than did the regulated tariffs. Particularly noteworthy were the increases for CSP (56.2%), wind (35.5%), hydro (22.6%), and biomass (8.5%). With one notable exception, increases in regulated tariffs grew by between 0 and five percent, and that for biomass actually fell by 5.2%. At the then prevailing TEM, most producers would earn several eurocents more per kilowatt-hour by opting for the market option.

The notable exception concerned solar PV. The premium option was eliminated for installations under 100 kW, but the cap for the highest level of support was raised from 5 to 100 kW. Thus installations with capacities in that range received an increase of more than 90% in the regulated tariff.

Finally, the new regulations included a number of measures to deal with the challenges to stability posed by incorporating a growing amount of intermittent power into the grid. Installations with a capacity of 10 MW or more that received the tariff were required to provide an output forecast at least 30 hours in advance to the grid operator, starting no later than 2005. Should actual generation deviate from the forecast by more than 20% for solar or wind, and five percent for other sources, a plant would be subject to a penalty, though corrections could be made until shortly before the opening of the market on a given day. This provision applied to all installations, including those built under both the 1994 and 1998 special regime rules. As a positive incentive, additional payments were available to installations that could contribute to supply stability, provide reactive power, and guarantee capacity (IDAE 2005, 50).

## *Impact of* Real Decreto *436/2004*

The new regulations went into effect in late March 2004, several weeks before the newly elected Socialist government could take office. In some respects, they were a success. From 2004 to 2006, wind capacity grew at the blistering pace of nearly 1900 MW per year, achieving 88% growth in just three years. It was on track to surpass the target of 13,000 MW as early as 2007. And with few exceptions, wind producers selected the market with premium option, in part to take advantage of rising electricity prices. Meanwhile, solar PV continued its exponential growth. Total capacity doubled in both 2004 and 2005 and then tripled in 2006. As a result, it too was poised to reach its admittedly much lower target of 150 MW in 2007 (del Río González 2008, 2925; Jacobs 2012, 145; Eikeland and Saeverud 2007, 28).

Other technologies, especially those that could be managed to some extent, did not fare nearly as well. The biggest disappointment was biomass, which was still expected to provide approximately as much electricity as wind in 2010. Its capacity grew by only 15 MW in 2004 and not much more thereafter. As a result, it had achieved only one-sixth of its capacity target by 2006. Minihydro capacity grew somewhat faster—by 234 MW over three years—but it was barely on track to reach its 2010 target of 2400 MW.

These setbacks, however, were due in no small part to factors beyond the scope of the regulations. The use of biomass to generate electricity, for example, was hindered by bottlenecks in collection and transport (IEA 2005, 125; see also IDAE 2005, 214–221). Likewise, the development of new hydroelectric capacity faced a litany of obstacles, including competing uses of waterways, the need to acquire water rights, environmental restrictions, and complicated administrative procedures. According to one estimate, obtaining approval for a new dam took an average of five years, and even then, municipalities often failed to issue the necessary work permits (IDAE 2005, 78–79, 80).

Nevertheless, some problems with *Real Decreto* 436/2004 were apparent from the outset. As a result, hardly had the text been published than it was subject to a torrent of criticism. Indeed, the Socialists promised to modify it even before they took office (APPA 2005a, 3). Other limitations would become evident with the passage of time. Thus, it would be only three years before it was replaced by another comprehensive set of regulations.

## Further Adjustments Under the Socialists: *Real Decreto* 661/2007

The period 2004–2007 saw yet another effort to get the details of the special regime right, this time under the new Zapatero government, which began to review renewable support policy soon after taking office. The already apparent shortcomings of *Real Decreto* 436/2004 were soon underscored by a further, generally upward, revision in the renewable capacity targets for 2010. This time, however, it took some two years to develop a new set of regulations in what at times was a highly contentious process. The resulting *Real Decreto* 661/2007 increased support levels while including new measures to control costs. Following its entry into force in June 2007, wind power continued to grow strongly, but the most notable consequence was an unexpected boom in solar PV. As a result, Spain would be forced to revise its renewable support policies yet again just a year later.

### *Motives for Further Adjustments*

The motives for yet another revision of Spain's renewable power support mechanism began with the problems with *Real Decreto* 436/2004. The renewable sector was particularly incensed by the new penalties for deviations from forecast production levels, which would apply retroactively to existing plants exercising the market option. Critics argued that this feature effectively negated the promise of a fixed price in the earlier regulations and, going forward, would deter developers of new projects from choosing the market option (APPA 2004a; IDAE 2005, 55).

From the beginning, moreover, there was also concern whether the support levels were high enough to meet the targets for 2010, notwithstanding the many increases in them. In particular, those for biomass and solar installations of more than 100 kWh, which had gone up the least, were regarded as inadequate. And for all the technologies, *Real Decreto* 436/2004 introduced uncertainty as to what would happen once the targets were reached (APPA 2004a; Dinica and Bechberger 2005, 275; IDAE 2005, 54, 166).

At the same time, some feared that the regulations did not do enough to control costs. They lacked any strong incentive for cost reduction, since the support levels could be reviewed by the government only every four years. In between, support would vary with the average cost of electricity, which jumped by 4.5% between 2005 and 2006. And producers could move back and forth between the two options in order to take advantage of whichever happened to be the most profitable at a given time (IEA 2005, 124).

Finally, *Real Decreto* 436/2004 did nothing to resolve enduring problems with grid access, connection, and capacity. The regulations on access and connection dated to 1985, and although the right to connect and priority access were enshrined in an EU directive and national law, they remained to be embodied and clarified in the relevant regulations. Also needed was a mechanism for determining who—producers or distributors—would bear the cost of reinforcing the grid in places where new renewable capacity threatened to overload it (APPA 2004c; IDAE 2005, 37, 52, 54; del Río González 2008, 2925; del Río and Mir-Artigues 2014, 39).

To make matters worse, it soon became clear that Spain would have to raise its targets for renewable capacity and power generation yet higher if the overall goals of 12% of primary energy consumption and 30% of electricity production from renewable sources were to be achieved. The principal culprit was rising demand for electricity, which increased by six percent per year from 1999 to 2005. As a result, demand already stood at more than 10% above the forecast for 2010 contained in the 1999 renewable energy plan (260.1 TWh), and it was expected to grow at two percent annually thereafter even with extensive efforts at energy efficiency and conservation. Partly as a result, the overall share of renewables in the energy mix stagnated or was even dropping (BP 2020; IDAE 1999, 37; IDAE 2005, 7–8, 325; APPA 2005a, 3).

At the same time, the growth in renewables was increasingly seen as critical for achieving Spain's climate change objectives. In 2004, renewable energy was included as a fundamental instrument for reducing $CO_2$ emissions in the first national plan for the assignment of emissions rights (*Plan Nacional de Asignación de Derechos de Emisión* 2005–2007 or PNA). To achieve its goals, the PNA contemplated an increase in renewables on the order of 22% over the targets contained in the 1999 plan (IDAE 2005, 8, 20, 23).

These pressures culminated in the government's approval in August 2005 of a new renewable energy plan that established a generally higher set of targets. With regard to the overall share of renewable generation in gross electricity production in 2010, the plan formally addressed the EU target for Spain by setting an even more ambitious national goal of 30.3%.

To achieve this goal, however, renewable output would have to reach 102 TWh, an increase of one-third over the target for 2010 contained in the 1999 plan. The share of renewables in overall primary energy consumption was also raised slightly, to 12.1%. If these targets were achieved, Spain would reduce its projected $CO_2$ emissions by 27.34 million tonnes overall and 18.65 million tonnes in the electricity sector, which would help to meet the country's Kyoto target of 333.2 million tonnes (IDAE 2005, 9, 32, 325, 335).

As for the targets for individual technologies, the biggest increase took place in that for wind, which jumped by more than 50%, from 13,000 MW to 20,155 MW, reflecting that technology's continuing rapid growth and seemingly boundless potential. Excluding hydro installations of more than 50 MW, wind would account for some 70% of all renewable capacity and nearly 60% of all renewable power generation in 2010. Also rising substantially in percentage terms were the targets for solar PV, from 150 to 400 MW (167%), and CSP, from 200 to 500 MW (150%). The former increase responded to PV's geometric rate of growth, while the latter reflected the large number of CSP projects either under construction or in development. Biogas also saw an increase in its target, from 78 to 235 MW (200%). Together, however, these three technologies would still contribute less than five percent of renewable electricity production in 2010 (IDAE 2005, 325, 327).

For the remaining technologies, the targets for 2010 were actually lowered. The most dramatic example was that of biomass, which saw its target decline by more than one-third from that in the 2002 planning document, from 3098 MW to just 2039 MW, reflecting the continuing obstacles encountered. To achieve even this lower target, moreover, nearly one-third of the biomass (722 MW) would have to be burned with coal, which would require a modification of the 1997 Electricity Sector Law. The target for minihydro was also reduced, though by a smaller percentage, from 2380 to 2199 MW. Nevertheless, biomass and minihydro would still be the second and third most important sources of renewable electricity, respectively, after wind power.

The new plan also estimated the costs of achieving these revised targets for renewable power. The total investment involved would amount to 18.99 billion euros, but only 0.25% of that (48.8 million euros) would be required in the form of government assistance. Instead, by far the largest public contribution would take the form of the premiums required to finance the additional generation. These would rise steadily over the implementation of the plan, for a total of 4.956 billion euros for the period 2005–2010 and 1.828 billion euros in 2010 alone. At that time, the average premium per kWh of electricity would be 4.3 eurocents (IDAE 2005, 275–276, 331–332).

Not surprisingly, more than half of the total investment (11.756 billion euros) and the additional premiums (2.599 billion euros) would go to new wind installations, but the average cost of the new wind power in 2010 would be only 3.1 eurocents per kWh. In contrast, the average premiums for PV and CSP in 2010 would amount to 36.3 eurocents per kWh and 19.6 eurocents per kWh, respectively. Because the output of these two technologies would be relatively low, however, the support costs for them (201 and 255 million euros, respectively) were not expected to impact the overall cost of the system dramatically. As a result, to cover the costs of the additional premiums, the electricity tariff would need to rise by only 0.6% per year (IDAE 2005, 275–276; APPA 2005a, 7).

### *The Development of New Regulations*

Despite the pressing need for yet another revision of the support system, the process of developing a new set of regulations took nearly two years. In late 2005, the Zapatero government announced that it would revise *Real Decreto* 436/2004 before the middle of the following year, but it did not submit a first draft to the National Energy Commission (*Comisión Nacional de Energía* or CNE), the regulatory body for the electricity sector, for review until late November 2006, after the government had approved the new plan. In the meantime, the renewable sector had offered detailed proposals for the future shape of the support system, and although some of these ideas were incorporated in the initial draft, the government's proposal was subject to a barrage of criticism (APPA 2005b, 2006a, 2006b).

The principal concern was that the proposed rules allowed for retroactive changes in the support levels. In response, dozens of companies and industry associations, regional governments, and other interest groups were motivated to comment on the draft. Then, in February 2007, the CNE issued an unfavorable judgment, emphasizing the importance of regulatory and juridical stability, and offered a long list of modifications. Its key proposal was that the new regulations should not apply to any installations put in operation before 2008 (CNE 2007).

Nevertheless, it is worth noting that the council of the CNE itself was deeply divided, with four of nine members, including the president, issuing a dissenting opinion. The crux of the dissent was that the majority's analysis had not taken into account the goal of minimizing the cost of the system, and that the current support scheme was generating unsustainably high rates of return. As we will see in the next chapter, these concerns would turn out to be well-founded.

### *Content of the New Regulations*

A final version of the new regulations, *Real Decreto* 661/2007, was issued in late May 2007 and went into effect at the beginning of June. It responded to the criticism by allowing existing plants to keep their current support levels or to opt into the new ones. And like the previous regulations, it sought to strike a balance between encouraging investment while avoiding high costs that would put upward pressure on electricity prices. Reflecting the growing complexity of doing so, the final document was some three times as long as its predecessor, filling 66 pages in the government's official bulletin.

The most innovative feature of *Real Decreto* 661/2007 was the introduction of a cap and floor system for generators choosing the market option. The floor guaranteed investors a minimum revenue stream and profit, no matter how low the price of electricity might fall. At the same time, the cap protected against windfall profits in the event that the market price spiked.

Several other measures were intended to promote investment by reducing uncertainty and risk. For the first time, the regulations enshrined the right of priority access and connection to the grid, and they called for the creation of a mechanism for sharing the costs of grid reinforcements between generators and distributors, though the details remained to be worked out. Henceforth, annual updates to support levels would be linked to the consumer price index rather than the more volatile average electricity price. Support would be guaranteed for the full lifetime of the installation. And with one exception (discussed below), support levels could be revised by the government only every four years, beginning in 2010, and any changes would not be retroactive. Looking ahead, the regulations called for the development of a new renewable energy plan for the period 2011–2020 in time to inform the first review of support levels.

At the same time, tariff and premium levels were increased in a number of categories. The premium option was eliminated for solar PV, but the technology continued to receive by far the highest regulated tariffs, and the tariff for installations of between 100 kW and 10 MW was increased by 81.7%, to nearly the level enjoyed by the smallest installations. The tariff for CSP was increased by 17.24%, and at the current electricity price, its compensation under the market option went up by 20.54%. Likewise, the tariff for minihydro grew by 13.16% and that for biogas increased by 15.95%. Perhaps most striking were the increases for biomass, which saw its tariff jump by as much as 130.5% and its compensation under the market option by as much as 45.56% at the existing electricity price, depending on the particular source. For the first time, moreover, biomass was given the option of co-combustion with coal in order to promote its use.

Notwithstanding these many assurances and incentives, *Real Decreto* 661/2007 also contained a number of cost control measures, in addition to the cap on compensation under the market option. It eliminated the market incentive established by *Real Decreto* 436/2004 for new plants and would end it for existing plants after 2012. Support levels would be reduced to 80% of the original level after a fixed number of years, ranging from as few as five to as many as 25, depending on the technology and size of the plant, regardless of how much support levels might have increased in the meantime because of inflation. And most significantly, it established a review mechanism geared to a new set of targets derived from those contained in the 2005 renewable energy plan. Once 85% of the target for a particular technology had been reached, the government would establish a maximum period of at least, but possibly no more than, one year, during which new installations would continue to receive the established premium or tariff. Once that period had expired, however, new plants would receive only the market price unless new support levels had been approved by the government.

The regulations also strengthened the existing measures intended to maintain the stability of the grid. Plants of more than 10 MW would have to be connected to a control center, which could determine their output remotely. The capacity of those installations receiving the regulated tariff that were obligated to provide an output forecast was reduced to 1 MW, and the tolerance threshold for deviations was reduced to five percent of output for all technologies. Finally, *Real Decreto* 661/2007 introduced the option of time-of-day differentiation for hydro, biomass, and biogas installations receiving the tariff, whereby generators could choose to receive a higher level of support for production during peak demand hours, in return for lower support levels during off-peak hours.

### Impact of Real Decreto 661/2007: The Boom in Solar PV

Just before the new regulations were issued, the renewable sector criticized the government for a lack of transparency and insufficient dialogue in the development process. When they finally appeared, however, the sector expressed general satisfaction with the result. Indeed, its leading lobbying group, the Renewable Energy Business Association (*Asociación de Empresas de Energías Renovables* or APPA), an umbrella organization of mainly independent producers that was formed in 1987, claimed that the regulations incorporated many of its suggestions (APPA 2007b).

In any case, the final version was certainly regarded as a big improvement over earlier drafts. It eliminated much of the regulatory risk for new investments and provided adequate transition periods for existing installations. Although reaching the targets for 2010 no longer seemed possible, given the slow development of most technologies up to that point, the industry regarded *Real Decreto* 661/2007 as providing sufficient stability to enable Spain to meet the recently issued EU target for 2020 of covering 20% of primary energy consumption with renewables (APPA 2007a, b).

If anything, however, the renewable sector underestimated the impact that *Real Decreto* 661/2007 would have. For the next several years, wind grew at an even faster rate, falling just two percent short of its target for 2010. The first CSP plants came online in 2007, and their total capacity grew geometrically thereafter, surpassing the target of 500 MW in 2010. In contrast, minihydro and biomass capacity barely budged between 2007 and 2010, reflecting the continued presence of the obstacles that had dogged their development from the beginning (Gómez et al. 2016, 436–437).

Most dramatic of all, however, was the subsequent explosion in the amount of solar PV. In 2007, total capacity more than quadrupled, to 544 MW, easily exceeding the trigger point of 371 MW contained in *Real Decreto* 661/2007. And in 2008, it jumped by an astonishing 2708 MW, with new plants coming online at an average rate of more than 500 MW per month between July and September. That year, nearly half of all new capacity worldwide was installed in Spain, which suddenly ranked second globally in total PV capacity. This startling growth came as a surprise even to the renewable sector, which as late as mid-July 2008 expected only 1000 MW to be installed between January and September of that year (IEA 2009, 99; del Río and Mir-Artigues 2014, 10; APPA 2008).

A parallel rapid increase took place in the amount of renewable power generation. Output from wind power almost doubled between 2006 and 2010, nearly reaching the target of 45,511 gigawatt-hours (GWh) set in the 2005 plan. And at 6401 GWh, generation from solar PV was more than 10 times what had been targeted in the most recent plan. Even minihydro achieved its target, thanks to an unusually high amount of precipitation in 2010. Only biomass fell well short of its goals.

In 2010, total renewable output reached 95,571 GWh, well beyond the original target of 76,600 GWh set in the 1999 plan but still short of the revised target of 102.3 TWh in the 2005 plan. Because of a decline in electricity consumption occasioned by the financial and economic crisis, however, the renewable share of overall electricity production actually exceeded the percentage target of 30.3%, although the figure could vary substantially from year to year because of variations in large hydroelectric output. And as we will see in the next chapter, this progress took place despite increasingly intense efforts to reign in the costs of the special regime.

### *Reasons for the Boom in Solar PV*

Much attention has been given to the reasons for the boom in solar PV in particular. del Río and Mir-Artigues have convincingly argued that it cannot be explained in terms of the tariff levels alone. It is true that the tariff for installations between 100 kW and 10 MW nearly doubled, but some 80% of the capacity installed in 2008 was in plants of less than 100 kW, which did not see a tariff increase. By that measure, the boom could just have easily begun under the previous regulations (Mir-Artigues 2012, 360–368; del Río and Mir-Artigues 2012, 5559–5561, 428–430; del Río and Mir-Artigues 2014, 12–14; Mir-Artigues and del Río 2016, 317–318).

Certainly, the high tariff levels made investment in PV increasingly attractive as the costs of installations steadily declined. Indeed, a drop in the price of silicon modules beginning in mid-2006 was reinforced by a strong euro, facilitating the massive import of solar panels. And developers figured out ways to game the system in order to increase the rate of return on investment from the intended 5–9% to as much as 10–15%.

But perhaps the most critical feature was the lack of an absolute cap on capacity or a degression mechanism for reducing the tariff automatically with increases in capacity. Instead, *Real Decreto* 661/2007 allowed new installations to continue to receive the full tariff for at least a year after 85% of the target for a given technology was achieved. In the case of solar PV, this occurred in the summer of 2007, just weeks after the new regulations went into effect. And for a variety of reasons, many developers and their investors decided to take advantage of this provision to rush new projects to completion rather than wait to see what level of support would be available a year or more later. Indeed, the first draft of a new regulation issued in September 2007 included a clear reduction in tariff levels, and in the event that a revised support system was not adopted, the support levels would fall to nothing.

Other factors supercharged this gold rush. A slowdown in housing construction generated a large pool of investment capital in search of profitable outlets. As a result, investors enjoyed easy access to credit on favorable terms. At the same time, the modular character of PV installations made it easy for small investors to get a piece of the action. Indeed, financial institutions and even the government heavily promoted investments in PV plants (e.g., IDAE 2008). This modularity and the fact that PV could be installed almost anywhere also facilitated its rapid deployment. Not least important, regional governments, hoping to reap the benefits of investment, offered additional economic incentives, simplified administrative procedures, and sometimes simply cut corners in order to facilitate the deployment of new plants.

## Conclusion

This chapter has traced the evolution of Spain's renewable power support schemes under successive governments and their impact, from the late 1990s to the late 2000s. This period saw a sustained effort to increase the amount of renewable power in order to reach increasingly ambitious goals. Overall, moreover, these efforts proved highly successful at promoting investment in renewable power. The total amount of renewable electricity generated rose rapidly and substantially, from 39.2 TWh in 1998 to 95.6 TWh in 2010.

At the level of individual technologies, however, a number of unexpected outcomes occurred. Perhaps the biggest, and most welcome, surprise was the rapid growth of wind power. Installed wind capacity increased from just 834 MW in 1998 to 19,702 MW in 2010, more than double the target of 8974 MW that had been set in 1999. In 2010, wind power generation was more than twice what had been projected in 1999 and, at 43 TWh, accounted for some 80% of the overall growth in renewable output. The success of wind, however, was tempered by the disappointing results in biomass, which had been expected to provide nearly two-thirds as much power as wind. Instead, its output in 2010 stood at less than one-quarter of its original target. Finally, there was the belated but dramatic growth in solar PV capacity, which leaped from almost nothing in 1998 and as little as 43 MW in 2005 to 3351 MW in 2008, many times the original target of 144 MW. And as the decade came to a close, CSP finally began to take off, passing its target of 500 MW in 2010. Nevertheless, the boom in solar PV in particular was a decidedly mixed blessing. It more than any other technology caused a sharp increase in the price tag of financing the special regime and set the stage for subsequent efforts to reign in those costs, the subject of the next chapter.

One striking feature of this period was the continued existence of significant cross-party support for promoting renewable power (see also del Río González 2008, 2917). As a result, governments from both sides of the political spectrum pursued very similar policies. Although the new special regime was first implemented and then refined by the People's Party governments elected in 1996 and 2000, it was maintained and further developed by the Socialist government that came to power in 2004.

Also noteworthy was the growing voice, and arguably the influence, of the renewable power sector. The decade saw a proliferation of associations of generators, equipment manufacturers, and developers, as APPA was joined by more specialized groups representing the wind industry, the solar PV industry, and other sectoral interests. These groups were afforded—and took advantage of—formal opportunities to comment on draft regulations in addition to whatever informal channels of influence they could muster and, indeed, appear to have played a growing role in shaping the regulations adopted during this period.

## Notes

1. The term "tariff" did not appear in the regulation.
2. 1 euro equaled 166.386 pesetas at the time of replacement (1999).

## References

APPA (Asociación de Empresas de Energías Renovables). 2004a. "Appa ante el nuevo Decreto de Retribución del Régimen Especial" (31 March). https://www.appa.es/notas-de-prensa-historico/ano-2004/.

———. 2004b. "Inicitiva legal de APPA para mejorar el RD 436" (28 May). https://www.appa.es/notas-de-prensa-historico/ano-2004/.

———. 2004c. "APPA plantea al Secretario General de Energía su propuesta de mejora del R.D. 436/2004" (15 June). https://www.appa.es/notas-de-prensa-historico/ano-2004/.

———. 2005a. *Info*, no. 20 (Nov.–Dec.). https://appa.es/wp-content/uploads/descargas/revista/APPAInfo20.pdf.

———. 2005b. "El Secretario General de la Energía manifiesta a APPA su intención de que las renovables tengan rentabilidad adecuada" (15 Dec.). https://www.appa.es/wp-content/uploads/2018/07/rentabilidadadecuadaDic05.pdf.

———. 2006a. "APPA solicita la incorporación del PER a la legislación vigente y algunas mejoras del marco establecido por el RD 436/04" (28 Nov.). https://www.appa.es/wp-content/uploads/2018/07/20061128-nuevo_regimen.pdf.

———. 2006b. "APPA rechaza en la CNE el grueso de la reforma del régimen económico de las renovables y propone alternativas válidas" (20 Dec.). https://www.appa.es/wp-content/uploads/2018/07/20061220-APPA-rechaza-en-la-CNEDic06.pdf.

———. 2007a. "Último llamamiento del sector al Gobierno para que no dé un paso atrás en el desarrollo de las renovables" (23 May). https://www.appa.es/wp-content/uploads/2018/07/20070523-APPA-AEE_El_sector_renovable_pide_al_Gobierno_que_no_de_un_paso_atras_May07.pdf.

———. 2007b. "APPA valora la nueva regulación de las energías renovables" (7 June). https://www.appa.es/wp-content/uploads/2018/07/20070607-APPA_valora_la_nueva_regulacion_de_las_energias_renovables_Jun07.pdf.

———. 2008. "Las propuestas de Industria paralizan el desarrollo del Sector Fotovoltaico español" (16 July). https://www.appa.es/wp-content/uploads/2018/07/20080716-Las_propuestas_de_Industria_paralizan_el_desarrollo_el_sector-fotovoltaico.pdf.

BP (British Petroleum). 2020. *Statistical Review of World Energy*. https://www.bp.com/en/global/corporate/energy-economics/statistical-review-of-world-energy.html.

CEC (Commission of the European Communities). 1996. *Energy for the Future: Renewable Sources of Energy: Green Paper for a Community Strategy.* COM(96)576 Final (20 Nov.). https://eur-lex.europa.eu/legal-content/EN/TXT/PDF/?uri=CELEX:51996DC0576&rid=2.

CNE (Comisión Nacional de Energía). 2007. *Informe 3/2007 de la CNE Relativo a la Propuesta de Real Decreto por el que Se Regula la Actividad de Producción de Energía Eléctrica en Régimen Especial y de Determinadas Instalaciones de Tecnologías Asimilables del Régimen Ordinario* (14 Feb.). https://www.cnmc.es/sites/default/files/1562569_8.pdf.

CNMC (Comisión Nacional de los Mercados y la Competencia). 2021. "Información mensual de estadísticas sobre las ventas de régimen especial. Contiene información hasta diciembre de 2020" (23 Feb.). https://www.cnmc.es/estadistica/informacion-mensual-de-estadisticas-sobre-las-ventas-de-regimen-especial-contiene-52.

del Río, Pablo, and Miguel A. Gaul. 2007. "An Integrated Assessment of the Feed-in Tariff System in Spain." *Energy Policy* 35, no 2 (Feb.): 994–1012.

del Río, Pablo, and Pere Mir-Artigues. 2012. "Support for Solar PV Deployment in Spain: Some Policy Lessons." *Renewable and Sustainable Energy Reviews* 16, no. 8 (Oct.): 5557–5566.

del Río, Pablo, and Pere Mir-Artigues. 2014. "A Cautionary Tale: Spain's Solar PV Investment Bubble." International Institute for Sustainable Development (Feb.). https://www.iisd.org/gsi/sites/default/files/rens_ct_spain.pdf.

del Río, Pablo, and Gregory Unruh. 2007. "Overcoming the Lock-Out of Renewable Energy Technologies in Spain: The Cases of Wind and Solar Electricity." *Renewable and Sustainable Energy Reviews* 11, no. 7 (Sept.): 1498–1513.

del Río González, Pablo. 2008. "Ten Years of Renewable Electricity Policies in Spain: An Analysis of Successive Feed-in Tariff Reforms." *Energy Policy* 36, no. 8 (Aug.): 2917–2929.

Dinica, Valentina, and Mischa Bechberger. 2005. "Spain." In *Handbook of Renewable Energies in the European Union*, edited by Danyel Reiche, 263–279. Frankfurt am Main: Peter Lang.

Directive 96/92/EC of the European Parliament and of the Council of 19 December 1996 Concerning Common Rules for the Internal Market in Electricity. *Official Journal of the European Communities* (30 Jan. 1997). https://eur-lex.europa.eu/legal-content/EN/TXT/PDF/?uri=CELEX:31996L0092&from=EN.

Directive 2001/77/EC of the European Parliament and of the Council of 27 September 2001 on the Promotion of Electricity Produced from Renewable Energy Sources in the Internal Electricity Market. *Official Journal of the European Communities* (27 Oct. 2001). https://eur-lex.europa.eu/legal-content/EN/TXT/PDF/?uri=CELEX:32001L0077&from=EN.

EC (European Commission). 1997. *Energy for the Future: Renewable Sources of Energy. White Paper for a Community Strategy and Action Plan.* COM(97)599 Final (28 Nov.). https://europa.eu/documents/comm/white_papers/pdf/com97_599_en.pdf.

Eikeland, Per Ove, and Ingvild Andreassen Saeverud. 2007. "Market Diffusion of New Renewable Energy in Europe: Explaining Front-Runner and Laggard Positions." *Energy & Environment* 18, no. 1: 13–36.

Gómez, Antonio, César Dopazo, and Norberto Fueyo. 2016. "The 'Cost of Not Doing' Energy Planning: The Spanish Energy Bubble." *Energy* 101 (15 April): 434–446.

IDAE (Instituto para la Diversificación y Ahorro de la Energía). 1999. *Plan de Fomento de las Energías Renovables en España* (Dec.). https://www.idae.es/uploads/documentos/documentos_4044_PFER2000-10_1999_1cd4b316.pdf.

———. 2005. *Plan de Energías Renovables en España (PER) 2005–2010* (Aug.). https://www.idae.es/publicaciones/plan-de-energias-renovables-en-espana-2005-2010.

———. 2008. *El Sol Puede Ser Suyo: Respuestas a Todas las Preguntas Clave sobre Energía Solar Fotovoltaica* (Nov.). https://www.idae.es/uploads/documentos/documentos_EL_SOL_PUEDE_SER_SUYO_-_FV_nov08_FINAL_12-01-2009_(2)_b6ef3ccd.pdf.

IEA (International Energy Agency). 2005. *Energy Policies of IEA Countries: Spain 2005 Review.* Paris: OECD.

———. 2009. *Energy Policies of IEA Countries: Spain 2009 Review.* Paris: OECD.

Jacobs, David. 2012. *Renewable Energy Policy Convergence in the EU: The Evolution of Feed-in Tariffs in Germany, Spain, and France.* London and New York: Routledge.

Ley 54/1997, de 27 de noviembre, del Sector Eléctrico (28 Nov. 1997). https://www.boe.es/buscar/doc.php?id=BOE-A-1997-25340.

MINECON (Ministerio de Economía). 2002. *Planificación de los Sectores de Electricidad y Gas: Desarrollo de las Redes de Transporte, 2002–2011* (13 Sept.). https://energia.gob.es/planificacion/Planificacionelectricidadygas/desarrollo2002-2011/Planificacion/portada-indice.pdf.

Mir-Artigues, Pere. 2012. *Economía de la generación eléctrica solar: La regulación fotovoltaica y solar termoelétrica en España.* Civitas/Thomson Reuters.

Mir-Artigues, Pere, and Pablo del Río. 2016. *The Economics and Policy of Solar Photovoltaic Generation.* New York: Springer.

OECD (Organization for Economic Cooperation and Development). 1999. *Spain: Regulatory Reform in the Electricity Industry.* http://www.oecd.org/regreform/sectors/2497385.pdf.

Real Decreto 2818/1998, de 23 de diciembre, sobre producción de energía eléctrica por instalaciones abastecidas por recursos o fuentes de energía renovables, residuos y cogeneración (30 Dec. 1998). https://www.boe.es/eli/es/rd/1998/12/23/2818.

Real Decreto 1663/2000, de 29 de septiembre, sobre conexión de instalaciones fotovoltaicas a la red de baja tension (30 Sept. 2000). https://www.boe.es/eli/es/rd/2000/09/29/1663.

Real Decreto 436/2004, de 12 de marzo, por el que se establece la metodología para la actualización y sistematización del régimen jurídico y económico de la actividad de producción de energía eléctrica en régimen especial (27 March 2004). https://www.boe.es/eli/es/rd/2004/03/12/436.

Real Decreto 661/2007, de 25 de mayo, por el que se regula la actividad de producción de energía eléctrica en régimen especial (26 May 2007). https://www.boe.es/eli/es/rd/2007/05/25/661/con.

CHAPTER 4

# The Dark Ages: Reponses to the Renewables Boom

## INTRODUCTION

The politics of renewable power in Spain shifted abruptly as the first decade of the 2000s came to an end. Where renewable power had previously enjoyed broad political support, it increasingly faced criticism and opposition from multiple actors and across the political spectrum. The immediate cause of this sea change was the sharp jump in support payments occasioned by the boom in solar photovoltaic (PV) and reinforced by continued growth in wind power production as well as the belated operation of the first concentrated solar power (CSP) plants. These costs then became associated with, and frequently blamed for, unsustainable growth in the so-called tariff deficit, which represented the difference between the revenues and outlays of the electric power system. The problem was exacerbated by a decline in the demand for electricity as a result of the financial and economic crisis, which made it difficult to increase revenues even as electricity rates were raised.

In response to this growing crisis, successive governments under both major political parties took a series of increasingly stringent actions, ushering in what might be regarded as the "Dark Ages" of renewable power in Spain. These measures can be grouped into three categories corresponding roughly to the chronological order in which they were introduced. First, both parties sought to limit the growth in renewable support costs by restricting the amount of new capacity that could be installed and the level of support new installations could receive. Second, and much more controversially, they made retroactive reductions in the support that could be received by existing installations while imposing new charges on electricity generators in order to raise additional revenue. Finally, in 2013 and 2014, the conservative government of Mariano Rajoy scrapped the special regime altogether and introduced a entirely new support system that would have the overall effect of further limiting the payments to renewable generators.

These actions had the intended effects as well as some important unintended consequences. The growth in renewable capacity slowed and then ground to a halt. At the same time, renewable support payments plateaued and then began to fall. These gains, however, came at considerable costs. One was a sharp drop in investment as well as employment in the renewable sector. More dramatically, many small investors in PV plants faced bankruptcy, as their revenues fell below the costs of servicing the loans they had taken out. They were forced to refinance their debts on less favorable terms because of the heightened risks now associated with such investments. And many investors decided to take the Spanish government to court in both domestic and international tribunals. The heady days of renewable power were over.

## MOTIVATIONS FOR REFORM: NEGATIVE CONSEQUENCES OF THE BOOM

Why did political support for renewable power vanish so dramatically? The explanation begins with the sharp rise in support payments that occurred concurrently with the implementation of *Real Decreto* 661/2007 and the boom in solar PV in particular. This sudden increase put substantial upward pressure on electricity costs just as the country headed into recession and electricity demand began to decline. As a result, revenues could not keep up with the costs of the electricity system, which incurred substantial deficits year after year.

### *Rising Costs of Support*

The 2005 renewable energy plan had estimated that total annual renewable power support costs would increase by 1838 million euros by 2010. And for the first few years, actual support costs adhered closely to this trajectory, rising by an average of 260 million euros per year. In 2008 and 2009, however, they jumped dramatically, by nearly 1000 million euros and then an eye-popping 3300, respectively. In 2010, renewable support costs stood at 5371 million euros, up some 4800 since 2004. At that time, the average cost of each unit of renewable electricity came to 7.8 eurocents per kilowatt-hour (kWh), and when spread out over all electricity consumed, support for renewable power still amounted to more than 2 eurocents/kwh because some one-quarter was coming from renewable sources (Mir-Artigues 2012, 360; del Río and Mir-Artigues 2014, 12, 50; CNMC 2013, Table 1.1; CEER 2013, 19/53).

The steady growth in wind power was partly to blame for this burden. Installed capacity had grown from 8500 MW in 2004 to 19,700 MW in 2010, and its equivalent support (after subtracting the wholesale electricity price) had grown in proportion, from 452 million euros to 1960 million euros, or nearly two and a half times the amount projected in the 2005 plan. In addition, the first CSP plants came online in 2007 and would account for an increasingly significant share of renewable support.

But the principal culprit in this tale was solar PV, whose equivalent support had grown from just 6.1 million euros in 2004 to 2650 in 2010, or some 50% of all renewable support costs. The reason was not that solar PV had come to account for a high percentage of all the renewable power produced; its share stood at around just 11%. Rather, it was PV's high equivalent support level per unit of power produced, which was roughly 10 times that of wind. During the early years of the special regime, this discrepancy did not much matter, because PV's contribution to renewable power generation was such a small share of the total. But with the boom in PV, its much higher support level began to be felt, even though wind still provided nearly seven times as much power. PV's share of renewable support was more than four times its share of renewable production (Mir-Artigues 2012, 360; del Río and Mir-Artigues 2014, 12).

### *Contribution to the Tariff Deficit*

Partly as a result of the sudden rise in renewable support costs, Spain experienced, beginning in 2008, a string of years in which the costs of the electricity system greatly exceeded its revenues. This so-called tariff deficit reached more than 6000 million in 2008 and then fluctuated between 3800 and 5600 million euros for the next several years, notwithstanding repeated substantial increases in the access tariff paid by customers that totaled 60% between 2007 and 2011. Eventually, the accumulated deficits reached nearly 30 billion euros, or approximately three percent of Spain's GDP, at a time when the government was under intense pressure to reduce the national debt[1] (Fig. 4.1).

Although renewable support costs contributed to the tariff deficit, they alone were not to blame. A more fundamental reason was the government's policy, in place since the liberalization of the power market in 1997, of controlling electricity prices to protect consumers against sudden increases, and a 2002 regulation temporarily capped annual increases at just two percent (Real Decreto 1432/2002). For that reason, Spain experienced its first substantial tariff deficit, of more than 4000 million euros, in 2005, when renewable support levels were still quite low but consumer tariffs reached their lowest point in real terms, just 70% of their 1996 value. Even as late as 2010, the cost of electricity to households was still lower than it had been when the restrictions had first been adopted, more than a dozen years before, despite relatively large increases during each

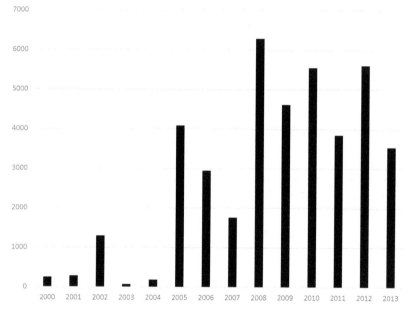

**Fig. 4.1** Tariff deficit, 2000–2013 (millions of euros) (*Source* Monforte 2020b)

of the previous several years (Mir-Artigues 2012, 416; del Río and Mir-Artigues 2014, 8–9). And notwithstanding those increases, beginning in 2008, total revenue was growing more slowly, if not declining, because of the drop in electricity consumption occasioned by the economic crisis.

Moreover, support payments for renewable power constituted only one of several regulated costs of the electric power system. Others included the costs of maintaining and expanding the transmission and distribution networks; subsidies for electricity systems on the islands, low-income households, energy efficiency, and the use of domestic coal; reimbursements for shuttered nuclear power plants; and payments on the pre-existing debt. Non-renewable power sources covered by the special regime, co-generation and solid waste, received substantial amounts of support as well (Espinosa 2013b; Gómez et al. 2016, 435; IEA 2015, 129; UNEF 2013).

Finally, it should be noted that the growth of renewables also exerted downward pressure on wholesale electricity prices by displacing other forms of electricity generation with higher marginal costs, especially fossil fuels. A substantial economic literature exists on the overall impact of renewables on electricity prices, and whether the cost of support was greater or less than this positive market effect. The earliest studies suggested that the net impact was in fact positive while later analyses found that the results varied over time and by technology, with wind yielding a net gain and solar PV a net loss (Sáenz et al. 2008; Ciarreta et al. 2014; Costa and Trujillo 2015; Carvalho and Pereira 2018).

In any case, these nuances tended to be lost in the debate over how to address the tariff deficit. The regulated costs of the electricity system were rising, by more than 200% between 2005 and 2013, and the largest single contribution to this increase—44% in 2013—was support for renewable power generation. As a result, this support received particular scrutiny, and high priority would have to be given to first limiting the increase of and then reducing those costs (IEA 2015, 99, 120; Espinosa 2013b).

*Political Implications*

One final consideration shaped the political response to the rapid growth of renewables in the late 2000s: the impact on other power generators. As noted above, the increase in renewable output put downward pressure on the wholesale price of electricity, reducing the revenues of other producers. To make matters worse, it contributed to overcapacity in the system, especially when demand for electricity began to drop in 2008. The average load factor in the power system fell from 44% in 2000 to 31% in 2012. As a result, conventional generators on the whole were selling less electricity at lower prices than would have been the case in the absence of so much renewable power on the market.

Particularly hard hit were the owners of natural gas-fired combined cycle (GFCC) power plants. In anticipation of rapid growth in electricity consumption, as forecast by the government, as well as pressure to reduce greenhouse gas emissions, utilities had invested heavily in GFCC in the first decade of the 2000s. Indeed, overall GFCC capacity grew even faster than that for wind, from nothing in 2001 to nearly 27 GW in 2012, making gas the single largest source of installed capacity. Initially, the power generated by these new plants enjoyed plenty of demand. But after 2008, the load factor for GFCC declined precipitously, from around 50% to less than 20% in 2012, and then continued to fall, almost to single digits in 2014. Revenues dropped accordingly (Moreno and Martínez-Val 2011; Mir-Artigues 2012, 414; Gómez et al. 2016, 434, 436, 444; IEA 2015, 95, 120).

The political implications of these developments were profound. The promotion of renewable power went from enjoying widespread support, or at least tolerance, to facing opposition from many corners. As Linares and Labandeira (2013, 2, 11–13) have noted, an increasing share of Spanish society came to view renewable energy as a source of trouble rather than a solution, although for a variety of reasons. Electricity consumers, especially large consumers, had previously been protected by limits on increases in the price of electricity, but as these caps were lifted in response to the growing tariff deficit, they began to turn against renewables. Electric utilities, for their part, had initially benefited from their investments in renewable power, especially wind farms, since the electricity produced always received prices higher than their marginal costs. But these benefits diminished as the wholesale price declined, along with demand for electricity from their fossil-fuel fired power plants.

Many of these critics were well-organized interest groups who were listened to by the Spanish government. Their efforts were aided by the fact that the renewable sector itself became divided, with promoters of different technologies advocating different solutions. Under the circumstances, with support payments and the tariff deficit growing, the major political parties had little choice but to respond by trying to rein in the costs of supporting renewable power. The next sections detail their efforts to do so.

## Efforts to Limit the Growth in Support Costs: To the Moratorium

The initial government efforts in response to the boom in PV and the growth of the tariff deficit focused on slowing the increase in renewable power support costs. As early as 2007, the Socialist government of José Luis Rodríguez Zapatero began developing a new regulation, eventually adopted in September 2008, specifically targeting solar PV, which was viewed as the main problem. It limited the size of the plants that could receive support, established quarterly quotas of new capacity, substantially reduced the size of the regulated tariff, limited the support period to 25 years, and added a mechanism for automatically reducing the tariff as conditions warranted. A 2010 regulation further reduced the size of the tariff for PV by applying "correction" factors of up to 55% beginning the following year. Other regulations extended similar measures to renewable power installations of other types, especially wind and CSP. Yet, unsatisfied with the results of these efforts and facing intense pressure from utilities and large electricity consumers, the subsequent conservative government of Mariano Rajoy took the drastic step of imposing an open-ended moratorium on support for new renewable power plants as one of its first actions, in January 2012.

### *Initial Efforts to Contain PV Support Costs*

Work on a replacement for *Real Decreto* 661/2007 began in earnest as early as September 2007, just three months after it went into effect, when the government announced that 85% of the 400 MW target for solar PV had already been reached. At that time, the government issued a draft regulation that raised the target to 1200 MW, but reduced the tariff for ground-mounted systems by 30% beginning in 2008, established a mechanism for reducing the tariff by five percent a year for all systems thereafter, and limited support to 25 years. The draft was roundly criticized by the renewable sector as well as the energy regulator, the National Energy Commission (*Comisión Nacional de Energía* or CNE), for potentially making retroactive changes in the existing regulations, which were expected to apply through at least September 2008 (CNE 2007; Mir-Artigues 2012, 365).

Distracted by the general election of March 2008, the government did not issue a second draft until July 2008. The new draft lacked the retroactive features of the previous one, but placed sharp limits on the amount of new capacity that could be installed beginning in 2009 to 300 MW per year, limited the maximum size of new PV installations to 10 MW, and reduced future tariffs by up to 35%, depending on the type of installation. The government estimated that these measures alone could save 415 million euros per year. The solar PV sector responded by proposing a higher annual cap of 600 MW, arguing that by maintaining sufficient volumes, tariffs could be reduced by up to 10% per year. By this time, moreover, the CNE had grown concerned about the delay in developing a tariff regime for the period after September 2008, which created uncertainty in the sector. The CNE also appeared to be more open to regulatory changes that might appear retroactive as long as they applied only to installations completed after the changes took effect (APPA 2008; CNE 2008; Mir-Artigues 2012, 354–355).

The government issued the final regulation, *Real Decreto* 1578/2008, in late September 2008. As expected, it dealt exclusively with solar PV, and it sharply reduced the tariff as well as the amount of capacity that could receive support. It also sought to promote smaller plants and distributed generation in the form of installations on buildings. To that end, it established a new classification scheme, distinguishing between rooftop (up to 2 MW) and ground-mounted (up to 10 MW, well below the previous 50 MW limit) systems, and it set annual caps for each type starting at just 267 MW and 133 MW, respectively, though as a concession to the sector, it provided for temporary increases in the annual quota for ground-mounted systems of 100 MW in 2009 and 60 MW in 2010. The new capacity would be allocated on the basis of quarterly calls (*convocatoria*) to plants in the order in which they were inscribed in a new pre-assignment registry. In the event that two or more plants registered at the same time, priority would be given to that with the smallest capacity, but registered plants would remain eligible for the next call. Multiple installations connecting to the grid at the same point would be treated as a single project, putting an end to the phenomenon of "solar orchards" that had enabled small investors to qualify for higher support levels while benefiting from the economies of scale of larger projects.

For the first time, the regulations established a mechanism for revising the tariff offered to new plants. Initially, the tariff was set at 34 eurocents per kWh for small rooftop installations (up to 20 kW) and 32 for others, which represented a drop of 23 to 27%, depending on the type and size of installation, from the support levels offered in *Real Decreto* 661/2007. Moreover, the tariff could change from quarter to quarter depending on what percentage of the quota had been met in the previous call. Under this system, tariffs could drop by as much as 10% per year, but they could also rise if less than 50% of the capacity cap was met. As a further innovation, the annual quotas would rise or fall in inverse relationship to tariff changes in the previous year. Support was limited to 25 years, although tariffs would still be adjusted annually in accordance with the consumer price index (CPI).

Finally, the new regulations included measures to discourage speculation. They required developers to make a deposit of 50 euros per kW for small rooftop projects and 500 euros per kW for larger installations. In addition, new plants would have to be completed within 12 months or risk losing support.

*Real Decreto* 1578/2008 dramatically slowed the growth of PV support costs, although it was developed before the government realized the full extent of the PV boom and its concomitant financial burden. As late as July 2008, the sector expected only 1000 MW to be installed in the first nine months of the year, and the government anticipated that PV support levels would reach just 915 million euros in 2009, well below the actual 2600 million euros. As it turned out, moreover, the capacity caps in the new regulations did not cover the last quarter of 2008, allowing even more capacity to be installed, albeit at the lower tariff (APPA 2008; CincoDías 2008; Mir-Artigues 2012, 372).

### *Additional Steps*

Not long into the implementation of *Real Decreto* 1578/2008, moreover, it became clear that even deeper reductions in PV support levels were possible. The price of solar modules was falling rapidly, and enough projects had pre-registered to fill the quota for an entire decade. Thus, even the prospect of a 10% cut per year was insufficient to dampen the enthusiasm of developers, and it seemed to make no sense to wait until 2012 for a thorough revision of PV tariff levels (Mir-Artigues 2012, 432).

The Zapatero government issued a preliminary justification for reducing the compensation of new PV plants in August 2010 and final regulations, in the form of *Real Decreto* 1565/2010, in November. Under the guise of establishing technical requirements called for in previous regulations, the new ones lowered support levels for projects qualifying under *Real Decreto* 1578/2008 by applying a series of correction factors. These ranged from 95% for small rooftop to 75% for large rooftop and 55% for ground-mounted installations. When combined with the previous reductions, this amounted to a 70% cut in tariff levels for ground-mounted systems since *Real Decreto* 661/2007. In addition, any existing PV plant that was modified substantially, such as by replacing its solar panels, would henceforth receive the new, lower tariff. Through this measure, the government expected to save an additional 263 million euros in support costs annually by 2014 (Real Decreto 1565/2010; ASIF, n.d., 22–27).

Beginning in 2009, the Zapatero government also adopted measures to reduce the growth in support costs for other renewable technologies as part of its overall strategy for tackling the tariff deficit. A measure taken that year, *Real Decreto-ley* 6/2009, extended the pre-assignment registry to the entire special regime, raising the prospect of comprehensive annual quotas like those already in place for PV. To register, projects would have to deposit a guarantee of 20 euros per kW (100 euros/kW for CSP), among other requirements, and they would have to be completed within three years in order to receive support payments.

### *The Moratorium and Its Impact*

The conservative Rajoy government, which took office in late 2011, carried these efforts to limit cost growth to their logical conclusion. As one of its first acts, in January 2012, it imposed an effective moratorium on support for new capacity. Unfinished projects that were already registered, including approximately 1200 MW of wind and 1600 MW of CSP, could proceed, but some 4500 MW of wind capacity under development was excluded. The measure noted that the moratorium was only temporary, but it would not be lifted until the problem of the tariff deficit had been resolved. Nevertheless, the government maintained that the existing capacity was sufficient to meet demand for the next years and, in the meantime, nothing would prevent the construction of new plants that sold directly to the wholesale market (Real Decreto-ley 1/2012; Mir-Artigues 2012, 461–465).

All these measures had a profound effect on the deployment of new renewable capacity. Solar PV was the first to feel the impact, given the prompt adoption of *Real Decreto* 1578/2008 in response to the boom. Under that measure, capacity growth paused in 2009 and then resumed at a much slower pace, approximately 300–500 MW per year, than had occurred during the boom, before coming to a standstill at 4650 MW in 2013. Prior to the moratorium, virtually all of the quarterly capacity quotas were filled, even as the tariff levels fell, and the awarded amount represented only one-tenth of all the projects that had been inscribed in the pre-assignment registry (ASIF, n.d., 9–10; Mir-Artigues 2012, 383 and 386).

The other main technologies were also impacted, although not as immediately. Wind capacity continued to rise; indeed, it grew at a record rate in 2009 before gradually slowing and then plateauing at approximately 23,000 MW in 2013. For its part, CSP, the first units of which had become operational only in 2007, increased at an accelerating rate for the next five years and then, too, abruptly came to a stop, at 2300 MW, in 2013 as the moratorium achieved full effect.

## Efforts to Reduce Existing Costs

Despite the success of the efforts to limit the growth of support costs, it soon became clear that additional steps would be needed. The tariff deficit continued to rise rapidly, despite substantial increases in the tariffs charged to customers. Indeed, retail electricity prices nearly doubled between 2005 and 2011 (del Río and Mir-Artigues 2014, 58). Given the lack of good alternatives, both Socialist and conservative governments turned to measures intended to reduce the amount of support going to existing installations.

### *Inspections*

One of the first steps was to ensure that all the PV plants receiving support under *Real Decreto* 661/2007 were eligible to do so. In particular, it was believed that some developers had falsified the date of initial operation in order to qualify for the higher tariffs before they expired in September 2008. *Real Decreto* 1578/2008 authorized the CNE to conduct inspections, which began at the end of the year. The task was daunting, given that some 28,000 plants had been built during the short time *Real Decreto* 661/2007 had been in effect (Mir-Artigues 2012, 397–404; ASIF, n.d., 21–22).

Nevertheless, the CNE inspected more than 4000 plants over the next year or so and found irregularities in some 80% of them. For example, all the promised solar panels may not have been installed, or technical equipment required for normal operation may have been lacking prior to the deadline. Still, the government hesitated to take any punitive action, such as withholding support payments or imposing penalties, while it waited for the approval of a new regulation that spelled out the procedures for confirming the proper registration of PV plants (Mir-Artigues 2012, 400–401).

A first draft had been issued in early 2009, but it was found to violate the rights of the regional governments, which had in many cases registered the suspect plants (CNE 2009). Thus the new regulation, *Real Decreto* 1003/2010, did not take effect until August 2010. It required plants to submit a number of documents to the CNE, and if they failed to do so, the CNE could suspend support payments, and any plant found not to be in order would have to return any support already received. For those potentially not in compliance, however, it offered an amnesty of sorts. If the plant renounced its registration within two months, it could keep the payments it had already received and then qualify for the lower, but still ample, tariff available under *Real Decreto* 1578/2008 going forward. To the government's surprise, fewer than 1000 installations totaling just 65 MW of capacity took up this option, resulting in an annual savings of just 17 million euros (Mir-Artigues 2012, 398–404).

Thus, beginning in 2011, the CNE began halting payments, and by the middle of the year, nearly 1300 plants out of just over 9000 inspected had been suspended. So many owners appealed, however, that the government restored their payments until their cases could be resolved. And in the end, the government determined that only about 90 MW of capacity was out of compliance, resulting in a savings of 51 million euros per year, or just two percent of the overall PV support costs (Mir-Artigues 2012, 404–405).

## Retroactive Cuts in Support

Even before the limited ability of the inspections to generate savings had become clear, the Zapatero government began eyeing a series of more controversial measures: retroactive modifications in the terms of compensation for pre-existing power plants. And once again, PV plants bore the brunt of the changes. A first step was taken with *Real Decreto* 1565/2010, which had reduced the tariffs available to new PV installations under *Real Decreto* 1578/2008 by as much as 45%. The regulation also extended the 25-year limit on compensation established in *Real Decreto* 1578/2008 to all pre-existing PV plants. Although this restriction was challenged—and upheld—in court, it did not represent a serious threat to the industry, since any plant was likely to be amortized by then, if not completely obsolete (ASIF, n.d., 24; Mir-Artigues 2012, 436; del Río and Mir-Artigues 2014, 22, 41–42).

Thus much more controversial was the next step: a restriction on the number of effective operating hours for which PV plants could receive support. Development of this measure began in the middle of 2010. The initial draft covered wind and CSP as well as PV, but the government struck a separate deal with the first two sectors that was completed in late 2010. It limited the number of equivalent hours of operation for which plants could receive remuneration, depending on the type of plant and, in the case of CSP, its energy storage capacity. In practice, this measure was not expected to have much of an impact, given the levels that were set. More consequential was a cut in the compensation for wind installations by 35%, which the government hoped would reduce support costs by as much as 1100 million euros (Real Decreto 1614/2010; Mir-Artigues 2012, 457–458; García and Mariz 2012, 55).

In the case of PV, the government originally hoped to reduce the amount of support by some 1000 million euros (40%) per year, though this target was later reduced to 790 million euros. It argued that such reductions were possible because the many plants built under *Real Decreto* 661/2007 were generating 30% more power than had been assumed when the tariff levels were set, and prices for solar panels had fallen so far that some plants were already replacing the original panels with new ones with greater peak potential (Mir-Artigues 2012, 438–440).

The final details were revealed in late December with the publication of *Real Decreto-ley* 14/2010. As expected, it limited the number of operating hours for which PV plants could receive support each year. The limits for plants installed under *Real Decreto* 661/2007 were particularly severe, ranging from 1250 hours for fixed solar panels to 1707 hours for those with two-axis tracking, or about a 15 to 25% reduction in the number of hours that had been receiving the tariff, and would apply through 2013. The limits for more recent plants ranged from 1232 to 2367 hours, depending on the type of installation and the geographical location, and were expected to have much less of an impact, although they were subject to no time limit. As a concession to the industry, the measure increased the number of years during which compensation could be received to 28. Overall, the government calculated that these limits would save 740 million euros per year, not far below its original target (Mir-Artigues 2012, 442–443; del Río and Mir-Artigues 2012, 5563; del Río and Mir-Artigues 2014, 18, 22–23).

Not surprisingly, the PV industry reacted negatively to the retroactive cuts. It estimated that, on average, revenues would fall by 30% for plants operating under *Real Decreto* 661/2007 and 10% for the rest. As a result, many owners would be unable to service the loans they had taken out and, in the best case, would have to refinance, typically at higher interest rates. Many investors could be ruined. The industry also argued that the measures made no economic sense, since they disfavored plants that made better use of the grid and those in more favorable locations (ASIF, n.d., 27–33; Mir-Artigues 2012, 442–443; APPA 2010).

Finally, there was the issue of the legality of such cuts. The Zapatero government made at least two somewhat inconsistent arguments in their defense. On the one hand, it did not consider the hourly restrictions retroactive, since regulatory measures were not the same as contracts. On the other hand, regulations could be changed retroactively, even if they resulted in future decreases in revenues, as long as they did not require that previous support payments be returned (Mir-Artigues 2012, 438–440, 444). Ultimately, the measures resulted in a number of lawsuits, which are discussed below.

After taking office in December 2011, the conservative Rajoy government picked up where the Socialists had left off by making further retroactive changes. In early 2013, it altered the mechanism for the annual updating of tariff levels, replacing the CPI with a more restrictive construct that was expected to result in smaller increases. It also eliminated the premium option for non-PV technologies, such as wind, that sold their output directly to the wholesale market. Henceforth, they would have to choose between receiving the less lucrative regulated tariff or the wholesale market price alone (Real Decreto-ley 2/2013; del Río and Mir-Artigues 2014, 20, 42).

### New Charges

To make matters worse, the reductions in support experienced primarily by PV producers but also by others were exacerbated by a series of new charges that were intended to generate additional revenue for the government and that put even more pressure on the bottom line. Under the Socialists, *Real Decreto-ley* 14/2010 introduced a grid access charge, to be paid by all electricity generators, of 0.5 euros per MWh, as required by the European Union. This charge was expected to reduce the net revenues of PV plants by an additional 0.2–0.3% per year while raising 150 million euros annually (ASIF, n.d., 27–33; Mir-Artigues 2012, 440). For its part, the subsequent conservative government, in December 2012, imposed an even more burdensome generation charge of seven percent (22% for hydroelectric) on all power production. This new tax was expected to raise 3.0 billion euros, although much of that would come from non-renewable generation (Ley 15/2012; Espinosa 2013b).

### Impact on System Support Costs

What impact did all of these measures have on the level of the renewable power support under the special regime? Without access to plant level data, it is impossible to differentiate the effects of efforts to limit the additional costs of new plants from those to reduce the costs of existing installations. But one can discern the overall impact.

First, these steps did halt the growth of payments for PV installations. After tripling in 2006, quadrupling in 2007, quintupling in 2008, and nearly tripling again in 2009, PV support leveled off at just over 2.6 billion euros per year and then declined modestly before stabilizing at around 2.5 billion euros. The overall amount of renewable support, however, continued to climb, reaching more than 6.7 billion euros in 2013. The principal reason was the continued growth in wind capacity and CSP, which did not completely stop until that year.

Meanwhile, in part because of the growth in renewable support costs but also due to the decline in electricity demand, the annual tariff deficit stubbornly refused to shrink. In 2009, the Zapatero government had set a goal of eliminating the tariff deficit by 2013 (Real Decreto-ley

6/2009). This goal proved to be elusive. In 2012, the first year of the Rajoy government, the deficit grew by another 4 to 5 billion euros, well above the target for the year of 1500 million euros, and this notwithstanding a 15% increase in the access tariffs paid by consumers, which now constituted more than half of a typical electricity bill. The accumulated debt had reached some 26 billion euros, and the forecast for 2013 was no better: another 4.5 billion euros, notwithstanding all the steps that had been taken to reduce it. By this time, moreover, both the European Union and the International Monetary Fund were recommending an overhaul of Spain's electricity regulation. Thus the stage was set for an even more far-reaching response by the conservatives: the elimination of the special regime altogether and its replacement with an entirely different support scheme for renewable power (Espinosa 2013a, 1–2; El País 2013; Robinson 2013, 3).

## The End of the Special Regime

The outlines of the new support system were spelled out by the Rajoy government in July 2013 in *Real Decreto-ley* 9/2013. The reform was part of a comprehensive set of steps intended to eliminate the expected tariff deficit. Consumer access tariffs would be increased another 900 million euros or approximately 3.2%. The central government budget would cover an additional 900 million euros. And the remaining 2700 million euros would be cut from the regulated costs, with approximately half of that coming from support for generation under the special regime (Robinson 2013, 3; Espinosa 2013a, 2–3).

To achieve the latter objective, the new decree-law eliminated the cornerstones of the special regime, *Real Decreto* 661/2007 and *Real Decreto* 1578/2008, as modified by *Real Decreto-ley* 6/2009, and revised relevant sections of the 1997 Electricity Sector Law. In their place, it laid the foundation for a new framework, the "specific remuneration regime," that was intended to ensure both security of supply and the financial stability of the electricity system while facilitating the participation of renewable generators in the market at the lowest possible cost. In particular, it replaced fixed tariffs and premiums with a new potential form of compensation (*retribución*) that would ostensibly enable renewable installations to compete on an equal footing with other technologies and receive a reasonable rate of return (*rentabilidad razonable*), but not more. The level of compensation would be based not on electricity generated, as before, but on the initial investment and operating costs of a typical installation that was efficiently and well managed over its useful lifetime, after accounting for the revenues received from sales to the market. The target rate of return was initially set at the average interest rate on 10-year government bonds plus three percent, or approximately 7.4% before taxes, and could be revised every six years. (Real Decreto-ley 9/2013; see also IEA 2015, 100–102, 129; Espinosa 2013a, 1–5; del Río 2016, 9–12; Mir-Artigues et al. 2018).

The details of the new arrangement were fleshed out over the following year in a new Electricity Sector Law, a new set of regulations concerning renewable electricity, and a lengthy ministerial order. The new law, *Ley 24/2013*, formally ended the long-standing distinction between the ordinary and special regimes. Henceforth, all renewable producers would sell their output directly to the wholesale market, although they would continue to enjoy priority grid access and dispatch in accordance with EU directives. Existing plants would henceforth be covered by the new regulations while new capacity would have to be authorized through a process of competitive contests (*concurrencia competitiva*), which remained to be spelled out in detail. The new regulations, *Real Decreto* 413/2014, regulated production and laid out the methodology for determining the amount of compensation a plant could receive. The complicated formula included a number of parameters, such as the technology, the year of initial operation, and the amount of installed capacity. Finally, the ministerial order, which was not issued until June 2014, assigned each installation to a standard plant type and provided the detailed values, such as initial investment and cost of operation, needed to complete the calculation for each type, in appendices that went on for some 1750 pages. These values were subject to revision every three years.

Perhaps needless to say, the renewables sector reacted vociferously to the new framework, especially the changes in the terms of support. It noted that the concept of reasonable profitability was highly arbitrary, and that after taxes, the rate of return might be no more than five percent and possibly less, depending on the assumed standard costs of investment and operation. Overall, the new cuts in support could total more than 2000 million euros per year, some 50% more than the government's estimate, and that would be on top of those already imposed by previous retroactive measures. Moreover, some plants might see no more support at all, since the new formula for remuneration took into account the amount of support already received. Facilities that had earned a return higher than the new "reasonable" rate would see their remuneration cut accordingly. As a result, many companies and individual would be unable to make payments on the loans they had taken out and thus be driven to the brink of bankruptcy.

More generally, the renewables industry complained that it had been shut out of the decision-making process and unfairly singled out for blame. Other actors in the electricity sector, especially the major utilities, were equally, if not more, responsible for the tariff deficit, yet were not subject to cuts to nearly the same extent, if at all. The measures also overlooked the contributions renewables had made to lowering the wholesale price of electricity, resulting in savings of as much as 32,500 billion euros so far by the industry's own estimate. By delinking compensation from output, moreover, the measures would result in greater inefficiency and make the electrical system less competitive. Not least important, by creating so much uncertainty about future returns, the new regime would cause investors to flee Spain, and not just from the energy sector (APPA 2013, 2014a, 2014b. 2014c; see also Espinosa 2013a, 2013b, 3–5; Barrero 2013).

### *Impact on Deployment and the Tariff Deficit*

What effect did the end of the special regime and the introduction of the new support system have on the renewable sector? The impact on generating capacity and, by extension, output covered by the system was minimal. The growth in capacity had already been halted in early 2012, although it took two years for all the projects in the pipeline to be completed and come online. For its part, output increased from about 68.3 GWh to 74.7 GWh between 2012 and 2014, peaking at 78.3 GWh in 2013, before stabilizing at about 70 GWh. This trend reflected the last additions of capacity as well as above average wind conditions in 2013.

Instead, perhaps as expected, the principal impact was on the amount of compensation paid to renewables. Total support after subtracting revenues from the wholesale market (*retribución específica*) dropped by more than 20% between 2013, when the new regulations went into effect, and 2014, from over 6.7 billion euros to under 5.3 billion euros, where it remained for the next two years. This drop corresponded closely to the government's predicted savings of 1.35 billion euros, although some of it can be attributed to the decline in output noted above (CNMC 2021).

And when combined with the earlier measures, the total cuts amounted to some 30% of prior support (APPA 2016).

## Long-Term Consequences of the Reforms

Beyond their intended effects, the reforms enacted by both Socialist and conservative governments between 2008 and 2014 in response to the unsustainable growth of renewable support costs and the tariff deficit had a number of additional, often political consequences. First, the cumulative effect of the cuts was to strain greatly the finances of previous investors in renewable generating capacity. Particularly hard hit were the small investors, some 60,000 or so, in solar PV plants, many of whom had taken out loans backed by their homes and other personal property. These investors suffered revenue cuts on the order of 40–50%, which made it difficult to service those loans. In the end, relatively few went bankrupt, but many were forced to refinance over longer terms and at higher interest rates, eliminating the profits they had hoped to earn. By one estimate, the internal rate of return for PV investments was cut from a range of 9.1–11.3% to just 0.3–3.3% (Anpier, n.d.; Mir-Artigues et al. 2018, 327–328).

Harder to measure but still potentially significant was the impact on future investment in renewables and the electricity sector more generally. Many observers felt that the reforms had dealt a mortal blow to investor confidence. Not only had the government made retroactive changes in the support system, but the new scheme allowed it to revise at its discretion the parameters for determining compensation every three years going forward. At a minimum, these measures would increase the risk premium investors would have to pay and thus reduce the attractiveness of future investment (del Río et al. 2015, 23; Mir-Artigues et al. 2018, 330).

To compound the government's problems, many investors and owners made legal challenges to the retroactive cuts in both domestic and international tribunals. In the latter alone, Spain faced claims totaling some 10 billion euros, which put at risk much of the savings the government had hoped to make. Domestic courts ruled repeatedly in favor of the government, but foreign challenges were usually successful. As of early 2020, international bodies had found in favor of the plaintiffs some 13 times, on the grounds that Spain had violated the Energy Charter Treaty, and awarded damages totaling more than one billion euros, and dozens of additional cases were pending (Monforte 2020a).

Finally, the cuts provoked responses that altered the political landscape in ways that could shape future political outcomes. In particular, the period saw the emergence of prominent new interest groups in the renewable energy sector. The Renewable Energy Foundation (*Fundación Energías Renovables* or FER), established in 2010, sought to create a forum free from corporate influence that could provide an independent perspective on the subject. The National Association of Photovoltaic Energy Producers (*Asociación Nacional de Productores de Energía Fotovoltaica* or ANPIER), also founded in 2010, was formed to represent and defend the many small and medium owners of PV plants as they were beginning to suffer from the first cuts. And in 2011, a number of organizations joined together to create the Spanish Photovoltaic Union (*Unión Española Fotovoltaica* or UNEF) so that the PV sector could speak with a single, more influential voice (Mir-Artigues 2012, 435).

## Conclusion

The period from the late 2000s to the mid-2010s constituted a Dark Ages of sorts for renewable power policy in Spain. During those years, support for both new and existing installations was steadily reduced. As a result, new deployments ground to a halt, and the owners of many operating plants saw their revenues shrink substantially. Some even faced financial ruin. The boom in renewables was over, and the future looked bleak.

These increasingly restrictive policies were pursued by Socialist and conservative governments alike. In contrast to the previous period, where both of the major parties strongly promoted renewable power, there was now a broad political consensus on the need to reduce levels of support. What had changed?

As in earlier years, the leading actor remained the core executive, which initiated all the major policy changes. But the milieu in which the central government operated had been transformed. The opportunities and external pressures to grow renewables remained, as did the potential benefits—security, economic, and environmental—of doing so. But these positive factors were now outweighed by the costs, as symbolized by the growing electricity tariff deficit, and the need to cut government spending following the financial crisis. In addition, powerful interest groups, such as the major utilities and large electricity consumers, which had previously tolerated, if not supported, the promotion of renewables began to suffer losses and became vocal critics.

In short, the previously favorable political winds had shifted dramatically against renewable power. There was little even the increasingly well-organized renewable sector could do to stem the tide of cuts in support. It would not be long, however, before it experienced another reversal of fortune, which is the subject of the next chapter.

## Note

1. Formally, the tariff deficit refers to the difference between the access tariffs paid by consumers and the costs of regulated activities in the system. The total retail price of electricity to customers was the sum of the access tariff and the wholesale price.

## References

Anpier (Asociación Nacional de Productores de Energía Fotovoltaica). n.d. "Retroactividad Zero." http://old.anpier.org/informacion-sectorial/retroactividad.

APPA (Asociación de Empresas de Energías Renovables). 2008. "Las propuestas de Industria paralizan el desarrollo del Sector Fotovoltaico español" (16 July). https://www.appa.es/wp-content/uploads/2018/07/20080716-Las_propuestas_de_Industria_paralizan_el_desarrollo_el_sector-fotovoltaico.pdf.

———. 2010. "Las medidas retroactivas del Ministerio de Industria provocarán la ruina de miles de inversores" (23 Dec.). https://www.appa.es/wp-content/uploads/2018/07/20101223-NP_MITYC_RUINA_MILES_INVERSORES_appa_mm.pdf.

———. 2013. "Las asociaciones de renovables denuncian un proceso de expropiación encubierta" (17 July). https://appa.es/wp-content/uploads/descargas/2013/NdP_%20renovables_denuncian_proceso_expropiacion_encubierta.pdf.

———. 2014a. "APPA denuncia el ensañamiento del Gobierno con las renovables" (4 Feb.). https://appa.es/wp-content/uploads/descargas/2014/140204_NP-APPA-denuncia-ensanamiento.pdf.

———. 2014b. "Las renovables denuncian la estrategia del Gobierno para acabar con el sector" (20 Feb.). https://appa.es/wp-content/uploads/descargas/2014/20140220_NdP_APPA.pdf.

———. 2014c. "El Gobierno paraliza el desarrollo del sector renovable y pone en peligro las instalaciones existentes" (10 June). https://appa.es/wp-content/uploads/descargas/2014/RD_RENOVABLES_10_junio_vf.pdf.

———. 2016. "APPA denuncia la contribución desproporcionada de las renovables en la resolución del déficit de tarifa" (6 April). https://www.appa.es/

appa-denuncia-la-contribucion-desproporcionada-de-las-renovables-en-la-res olucion-del-deficit-de-tarifa-en-relacion-a-otros-costes-del-sistema-electrico/.

ASIF (Asociación de la Industria Fotovoltaica). n.d. *Hacia el Crecimiento Sostenido de la Fotovoltaica en España: Informe Anual 2011.* https://doc player.es/71781388-Hacia-el-crecimiento-sostenido-de-la-fotovoltaica-en-esp ana.html.

Barrero, Antonio. 2013. "El sector, a la espera de la letra pequeña de la 'rentabilidad razonable.'" *Renewable Energy Magazine* (17 July). https://www.renewableenergymagazine.com/panorama/respaldo--que-respaldo--201 30719/articulo/el-sector-a-la-espera-de-la-20130717.

Carvalho Figueiredo, Nuno, and Patrìcia Pereira da Silva. 2018. "The Price of Wind Power Generation in Iberia and the Merit-Order Effect." *International Journal of Sustainable Energy Planning and Management* 15 (15 Jan.): 21–30.

Ciarreta, Aitor, Maria Paz Espinosa, and Cristina Pisarro-Irizar. 2014. "Is Green Energy Expensive? Empirical Evidence from the Spanish Electricity Market." *Energy Policy* 69 (June): 205–215.

CincoDías. 2008. "Industria remite a la CNE el decreto fotovoltaico, que reduce hasta un 35% las primas" (21 July). https://cincodias.elpais.com/cincodias/2008/07/21/empresas/1216807175_850215.html.

CNE (Comisión Nacional de Energía). 2007. *Informe 31/2007 de la CNE Relativo a la Propuesta de Real Decreto de Retribución de la Actividad de Producción de Energía Eléctrica Mediante Tecnología Solar Fotovoltaica Para Instalaciones Posteriores a la Fecha Límite de Mantenimiento de la Retribución del Real Decreto 661/2007, de 25 de Mayo, Para Dicha Tecnología* (13 Dec.). https://www.cnmc.es/sites/default/files/1564817_8.pdf.

———. 2008. *Informe 30/2008 de la CNE en Relacion con la Propuesta de Real Decreto de Retribución de la Actividad de Producción de Energía Eléctrica Mediante Tecnología Solar Fotovoltaica Para Instalaciones Posteriores a la Fecha Límite de Mantenimiento de la Retribución del Real Decreto 661/2007, de 25 de Mayo, Para Dicha Tecnología* (29 July). https://www.cnmc.es/sites/def ault/files/1561483_8.pdf.

———. 2009. *Informe 10/2009 de la CNE sobre la Propuesta de Real Decreto por el que se Regula la Liquidación de la Prima Equivalente a las Instalaciones de Producción de Energía Eléctrica de Tecnología Fotovoltaica en Régimen Especial* (13 May). https://www.cnmc.es/sites/default/files/1559276_8.pdf.

CNMC (Comisión Nacional de los Mercados y la Competencia). 2013. *Información mensual de estadísticas sobre las ventas de régimen especial. Contiene información hasta diciembre de 2013* (1 Dec.). https://www.cnmc.es/estadi sticas?hidtipo=12749.

———. 2021. *Información mensual de estadísticas sobre las ventas de régimen especial. Contiene información hasta diciembre de 2020* (23 Feb.). https://

www.cnmc.es/estadistica/informacion-mensual-de-estadisticas-sobre-las-ven tas-de-regimen-especial-contiene-52.
Costa-Campi, Maria Teresa, and Elisa Trujillo-Baute. 2015. "Retail Price Effects of Feed-in Tariff Regulation." *Energy Economics* 51 (Sept.): 157–165.
CEER (Council of European Energy Regulators). 2013. *Status Review of Renewable and Energy Efficiency Support Schemes in Europe* (25 June). https://www.ceer.eu/documents/104400/-/-/fb65b156-71a6-b717-a3ec-585b138aa3ae.
del Río, Pablo. 2016. "Implementation of Auctions for Renewable Energy Support in Spain." Report D7.1-ES (March). http://auresproject.eu/sites/aures.eu/files/media/documents/wp7_-_case_study_report_spain_1.pdf.
del Río, Pablo, Anxo Calvo-Silvosa, and Guillermo Iglesias. 2015. "The New Renewable Electricity Support Scheme in Spain: A Comment." *Renewable Energy Law and Policy Review* 6, no. 1: 17–24.
del Río, Pablo, and Pere Mir-Artigues. 2012. "Support for solar PV deployment in Spain: Some policy lessons." *Renewable and Sustainable Energy Reviews* 16, no. 8 (Oct.): 5557–5566.
del Río, Pablo, and Pere Mir-Artigues. 2014. "A Cautionary Tale: Spain's Solar PV Investment Bubble." International Institute for Sustainable Development (Feb.). https://www.iisd.org/gsi/sites/default/files/rens_ct_spain.pdf.
El País. 2013. "El déficit de tarifa eléctrica desborda las previsiones con 4.281 millones" (18 Feb.). https://elpais.com/economia/2013/02/18/actualidad/1361186235_286878.html.
Espinosa, Maria Paz. 2013a. "An Austerity-Driven Energy Reform" (Oct.). https://www.researchgate.net/publication/258809913_An_austerity-driven_energy_reform.
Espinosa, Maria Paz. 2013b. "Understanding Tariff Deficit and Its Challenges" (March). https://www.researchgate.net/publication/258809901_Understanding_Tariff_Deficit_and_Its_Challenges.
García, Maria Teresa, and Rosa Mariá Mariz. 2012. "Analysis of the success of feed-in tariff for renewable energy promotion mechanism in the EU: lessons from Germany and Spain." *Procedia—Social and Behavioral Sciences* 65 (3 Dec.): 52–57.
Gómez, Antonio, César Dopazo, and Norberto Fueyo. 2016. "The 'Cost of Not Doing' Energy Planning: The Spanish Energy Bubble." *Energy* 101 (15 April): 434–446.
IEA (International Energy Agency). 2015. *Energy Policies of IEA Countries: Spain 2015 Review*. Paris: OECD
Ley 15/2012, de 27 de diciembre, de medidas fiscales para la sostenibilidad energética (27 Dec. 2012). https://www.boe.es/eli/es/l/2012/12/27/15.
Ley 24/2013, de 26 de diciembre, del Sector Eléctrico (26 Dec. 2013). https://www.boe.es/eli/es/l/2013/12/26/24.

Linares, Pedro, and Xavier Labandeira. 2013. "Renewable Electricity Support in Spain: A Natural Policy Experiment". *Economics for Energy WP 04/2013*. https://repositorio.comillas.edu/rest/bitstreams/18834/retrieve.

Mir-Artigues, Pere. 2012. *Economía de la generación eléctrica solar: La regulación fotovoltaica y solar termoelétrica en España*. Civitas/Thomson Reuters.

Mir-Artigues, Pere, Emilio Cerdá, and Pablo del Río. 2018. "Analysing the Economic Impact of the New Renewable Electricity Support Scheme on Solar PV Plants in Spain." *Energy Policy* 114 (March): 323–331.

Monforte, Carmen. 2020a. "España pierde un nuevo laudo, el número 13, por el recorte a las renovables." *CincoDías* (24 Jan.). https://cincodias.elpais.com/cincodias/2020/01/23/companias/1579783758_206659.html.

———. 2020b. "El sistema eléctrico registró un déficit de tarifa de unos 700 millones en 2019," *CincoDías* (7 Sept.). https://cincodias.elpais.com/cincodias/2020/09/04/companias/1599254601_996528.html.

Moreno, Fermín, and José Martínez-Val. 2011. "Collateral Effects of Renewable Energies Deployment in Spain: Impact on Thermal Power Plants Performance and Management." *Energy Policy* 39, no. 10 (Oct.): 6561–6574.

Real Decreto 1432/2002, de 27 de diciembre, por el que se establece la metodología para la aprobación o modificación de la tarifa eléctrica media o de referencia y se modifican algunos artículos del Real Decreto 2017/1997, de 26 de diciembre, por el que se organiza y regula el procedimiento de liquidación de los costes de transporte, distribución y comercialización a tarifa, de los costes permanentes del sistema y de los costes de diversificación y seguridad de abastecimiento (31 Dec. 2002). https://www.boe.es/eli/es/rd/2002/12/27/1432.

Real Decreto 661/2007, de 25 de mayo, por el que se regula la actividad de producción de energía eléctrica en régimen especial (1 June 2007). https://www.boe.es/eli/es/rd/2007/05/25/661/con.

Real Decreto 1578/2008, de 26 de septiembre, de retribución de la actividad de producción de energía eléctrica mediante tecnología solar fotovoltaica para instalaciones posteriores a la fecha límite de mantenimiento de la retribución del Real Decreto 661/2007, de 25 de mayo, para dicha tecnología (27 Sept. 2008). https://www.boe.es/eli/es/rd/2008/09/26/1578.

Real Decreto 1003/2010, de 5 de agosto, por el que se regula la liquidación de la prima equivalente a las instalaciones de producción de energía eléctrica de tecnología fotovoltaica en régimen especial (6 Aug. 2010). https://www.boe.es/eli/es/rd/2010/08/05/1003.

Real Decreto 1565/2010, de 19 de noviembre, por el que se regulan y modifican determinados aspectos relativos a la actividad de producción de energía eléctrica en régimen especial (23 Nov. 2010). https://www.boe.es/eli/es/rd/2010/11/19/1565.

Real Decreto 1614/2010, de 7 de diciembre, por el que se regulan y modifican determinados aspectos relativos a la actividad de producción de energía eléctrica a partir de tecnologías solar termoeléctrica y eólica (8 Dec. 2010). https://www.boe.es/eli/es/rd/2010/12/07/1614.

Real Decreto 413/2014, de 6 de junio, por el que se regula la actividad de producción de energía eléctrica a partir de fuentes de energía renovables, cogeneración y residuos (10 June 2014). https://www.boe.es/eli/es/rd/2014/06/06/413.

Real Decreto-ley 6/2009, de 30 de abril, por el que se adoptan determinadas medidas en el sector energético y se aprueba el bono social (7 May 2009). https://www.boe.es/eli/es/rdl/2009/04/30/6.

Real Decreto-ley 14/2010, de 23 de diciembre, por el que se establecen medidas urgentes para la corrección del déficit tarifario del sector eléctrico (24 Dec. 2010). https://www.boe.es/eli/es/rdl/2010/12/23/14.

Real Decreto-ley 1/2012, de 27 de enero, por el que se procede a la suspensión de los procedimientos de preasignación de retribución y a la supresión de los incentivos económicos para nuevas instalaciones de producción de energía eléctrica a partir de cogeneración, fuentes de energía renovables y residuos (28 Jan. 2012). https://www.boe.es/eli/es/rdl/2012/01/27/1.

Real Decreto-ley 2/2013, de 1 de febrero, de medidas urgentes en el sistema eléctrico y en el sector financiero (2 Feb. 2013). https://www.boe.es/eli/es/rdl/2013/02/01/2.

Real Decreto-ley 9/2013, de 12 de julio, por el que se adoptan medidas urgentes para garantizar la estabilidad financiera del sistema eléctrico (13 July 2013). https://www.boe.es/eli/es/rdl/2013/07/12/9.

Robinson, David. 2013. *Pulling the Plug on Renewable Power in Spain* (Dec.). https://www.oxfordenergy.org/publications/pulling-the-plug-on-renewable-power-in-spain/.

Sáenz de Miera, Gonzalo, Pablo del Río, and Ignacio Vizcaíno. 2008. "Analysing the Impact of Renewable Electricity Support Schemes on Power Prices: The Case of Wind Electricity in Spain." *Energy Policy* 36, no. 9 (Sept.): 3345–3359.

UNEF (Unión Española Fotovoltaica). 2013. "Una radiografía a la generación del déficit de tarifa" (5 July). https://www.energias-renovables.com/fotovoltaica/unef-radiografia-la-generacion-de-la-deuda-20130709.

CHAPTER 5

# The Renaissance of Renewable Power

## INTRODUCTION

The politics of renewable power underwent yet another reversal in the second half of the 2010s, when it experienced a Renaissance of sorts. The halt in deployment imposed by the center-right government of Mariano Rajoy proved to be only temporary. Indeed, hardly had the new regulatory system been put in place than pressure began to build for a further increase in Spain's renewable power capacity in order to achieve the country's overall renewable energy target for 2020. This pressure spurred the government to hold of a series of auctions through which it ultimately awarded rights to some 8700 megawatts (MW) of new renewable capacity, most of which could potentially receive remuneration under the new investment-based support system adopted in 2013 and 2014. At the same time, declining costs, especially for solar photovoltaic (PV) equipment, and new mechanisms for financing projects meant that Spain was reaching the point where government support might no longer be needed. As a result, the country experienced a second renewable power boom, which culminated in a record deployment of some 6.5 gigawatts (GW) of new capacity in 2019 alone.

© The Author(s), under exclusive license to Springer Nature
Switzerland AG 2021
J. S. Duffield, *Making Renewable Electricity Policy in Spain*,
Environmental Politics and Theory,
https://doi.org/10.1007/978-3-030-75641-3_5

## Pressures for Further Government Action

Even as successive Spanish governments sought to limit the growth of and then reduce the support costs of renewable power, they were adopting new, higher targets for renewable capacity and output. In 2009, the European Union (EU) established a target for Spain of 20% of energy consumption from renewable sources by 2020, meaning that well over 30% of the country's electricity would have to be renewable. The successive center-left and center-right governments translated this broad target into a series of renewable energy plans that called for substantial increases in generating capacity. Initially, the Rajoy government insisted that the moratorium would not impede Spain's ability to achieve these targets. By 2015, however, it became clear that additional efforts to promote renewable power would be needed if Spain were to satisfy its EU obligations. The principal remaining question was how to do so at the lowest possible cost.

### *Developments at the EU Level: Targets for 2020*

In the late 2000s, the EU moved to establish higher targets for renewable energy as part of a broader, integrated policy on energy and climate. The new targets would replace those set in the late 1990s and early 2000s. In March 2007, the European Council adopted a comprehensive energy Action Plan that, among other things, endorsed a binding target of a 20% share of renewable energy in overall EU energy consumption by 2020 (Council of the European Union 2007). Over the next two years, the EU worked out differentiated national targets, which were adopted in April 2009. Under the new EU renewable energy directive, Spain was required to provide at least 20% of its gross final energy consumption by 2020 from renewables. This target represented a substantial increase over the previous EU target of 12% set in 1997 (see Chapter 3). In addition, each member state was required to adopt a national renewable energy action plan that established sectoral targets and identified means to achieve them (Directive 2009/28/EC).[1]

### *The Zapatero Government's Response: New Renewable Energy Plans*

Over the next year, the Socialist government of José Luis Zapatero developed the required action plan, which was published in June 2010. The new plan set an even higher target for the renewable share of gross final energy consumption—22.7%—than that required by the EU and a correspondingly high share—38.2%—of electricity production. The latter was well above both the 2010 target of 29.4% and the actual level of 24% reached in 2009. To achieve these targets, Spain would have to increase substantially its renewable generating capacity between 2010 and 2020. Most notably, wind would have to grow by 17,845 MW (88.5%), concentrated solar power (CSP) by 4447 MW (704%), and PV by 4346 MW (108%). Two-thirds of the new PV would be smaller rooftop systems, consistent with the implementation of *Real Decreto* 1578/2008. Overall, renewable generating capacity would rise by 28 GW (67.5%), to a total of nearly 70 GW, and output by almost 66 TWh (78.5%) (IDAE 2010, 16–17, 39, 45, 159–160).

It should be noted that these figures were based on the assumption that energy efficiency measures beyond those already planned would reduce primary energy demand by 11% (IDAE 2010, 27–28). If this assumption proved to be wrong, capacity increases would also certainly have to be even greater to achieve the EU target. As ambitious as these targets were, moreover, they represented a reduction from a previous draft of approximately seven percent for total renewable capacity, from 74,547 MW, and of some 14%, from 15,695 MW, for PV and CSP combined. Critics blamed the cuts on the influence of the major utilities, although the government itself attributed them to the potential impact on the cost of electricity (Energías Renovables 2010a, 2010b).

Over the next year and a half, these targets were lowered somewhat more, in two further planning exercises, but still remained substantial. The first was a cross-party congressional analysis of Spanish energy strategy for the next 25 years, which began in mid-2009. The final report, issued in December 2010, reduced the overall targets for the renewable share of final energy consumption to 20.8% and of gross electricity generation to 35.5%, which were still well above the existing levels. Nevertheless, renewable electricity production would nearly double, from 72.8 TWh to 139.6 TWh, thanks to a nearly 25 GW increase in renewable capacity over the 2009 level of 39.5 GW, although no breakdown by technology was provided (Congreso de los Diputados 2010, 73–75, 119).

Then, in late 2011, just before the general election, the Zapatero government approved a new renewable energy plan, as called for in *Real Decreto* 661/2007. Over 800 pages in length, the plan provided a much more detailed analysis than was possible in the 2010 action

**Table 5.1** Renewable power targets for 2020

|  | 2010 | 2020 Targets | |
| --- | --- | --- | --- |
| Capacity (MW) | Actual | 2010 | 2011 |
| Wind | 20,744 | 38,000 | 35,750 |
| PV | 3787 | 8367 | 7250 |
| CSP | 632 | 5079 | 4800 |
| Biomass | 533 | 1187 | 1350 |
| Generation (GWh) | | | |
| Wind | 43,708 | 78,254 | 72,556 |
| PV | 6279 | 14,316 | 12,356 |
| CSP | 691 | 15,353 | 14,379 |
| Biomass | 2820 | 7400 | 8100 |

*Notes* Actual values for 2010 and 2010 targets for 2020 are from IDAE (2010, pp. 159–160). 2011 targets for 2020 are from IDAE (2011a, pp. 469–470). Capacity targets for wind include 3000 MW of offshore wind in 2010 and 750 MW in 2011

plan. It adopted the overall target for renewables found in the congressional report—20.8%—but raised the goal for gross electricity production backup to 38.1%. The plan called for substantial increases in renewable generating capacity, although not as high as those in the action plan. Wind would grow to 35,750 MW (2250 MW less, all in offshore), PV to 7250 MW (1117 MW less), and CSP to 4800 MW (279 MW less). Only the target for biomass (1750 MW) was increased, by a modest 163 MW (IDAE 2011a, 469–475; 2011b, 6–7, 26) (Table 5.1).

The lower targets reflected the growing pressure to reduce support costs. Even then, it was estimated that the new capacity would receive an additional 23.2 billion euros in support over 10 years, although the extra cost was expected to peak in 2014 and then decline steadily thereafter as a share of the total costs of the system (IDAE 2011a, 570, 590; del Río 2016, 7).

### *Initial Reactions by the Rajoy Government*

Ironically, the adoption of the plan was almost immediately followed by the moratorium on the awarding of support to new projects that was imposed by the newly elected conservative government in early 2012. Initially, the Rajoy government maintained that the moratorium would not jeopardize the achievement of the 2020 targets. Indeed, the implementing legislation claimed that "the generation capacity installed at this time is sufficient to ensure the coverage of the expected demand for the coming years" and "(T)hus, it is not essential at this time to continue with the annual rates of implementation of these technologies to achieve the expected objectives at the end of the decade" (Real Decreto-ley 1/2012).

It was not long, however, before concerns began to be raised. As early as May 2012, an EU commission working document noted that suspending support would make it difficult to achieve Spain's 2020 renewable energy target. By 2015, Spain was clearly falling short of its own goals, and that year's EU renewable energy progress report concluded that the country would need to reassess the adequacy of its policies (EC 2012, 23; 2015, 5; Planelles 2017).

Indeed, the government's own planning process was already arriving at the conclusion that more renewable power would be required. An analysis of the needs of the electricity and gas sectors through 2020, completed in 2015 and approved by the government's Council of Ministers that October, anticipated an increase in capacity of as much as 8537 MW above the 2013 level, including 6473 MW of wind and 1370 MW of PV (MINETUR 2015b, 28).

Political developments also militated in favor of renewed action. The Rajoy government faced national elections before the end of the year, and it was eager to demonstrate that it supported investment in renewable energy. Indeed, the party's electoral platform devoted three planks to the subject (Pérez Rodriquez 2015; del Río 2016, 12 and 26; Partido Popular 2015, 37–38).

## THE RAJOY GOVERNMENT'S RESPONSE: THE AUCTIONS

The next question was how best to go about promoting the deployment of additional renewable generating capacity. In fact, much of the necessary groundwork had been laid during the previous two years, both in Spain and within the EU.

At the EU level, this preparation took the form of new guidance for the design of renewable support schemes. A 2013 draft concluded that "For renewable electricity…well-designed auction systems should provide the most cost-efficient conditions for delivering renewables…. Tendering for the desired volume of energy… is the most economically efficient means of delivering this goal [of reducing costs]" (EC 2013, 7). Likewise, the final guidelines on state aid for energy, issued in 2014, extolled market instruments as a means of reducing subsidies to a minimum and required that, beginning in 2017 and with only limited exceptions, all aid be "granted in a competitive bidding process on the basis of clear, transparent and non-discriminatory criteria" (EC 2014, 24 and 26).

Preliminary steps had also been taken in the Spanish electricity sector reform of 2013–2014. The new law (Ley 24/2013 (Art. 14.7)) provided for "competitive contests" to provide support for new installations when needed to meet EU obligations or to reduce the cost of energy or foreign energy dependence. For its part, the implementing regulation for renewable energy, *Real Decreto* 413/2014, simply noted that a further Royal Decree and Ministerial Order would be issued to establish the parameters of any future competitive contest. It was not until the following year, as pressure grew for a further increase Spain's renewable capacity, however, that the details of the first auction were worked out.

## *The 2016 Auction*[2]

The first auction was held in January 2016. First drafts of the needed regulations were submitted to the national energy regulatory body (*Comisión Nacional de los Mercados y la Competencia* or CNMC) for review in April 2015, and final versions (Real Decreto 947/2015 and Orden IET/2212/2015) were issued the following October, shortly before the national election. The scope of the auction was intentionally modest, reflecting the government's lack of experience with such matters. It would be limited to just two technologies, wind and biomass, and relatively small amounts of capacity—500 MW and 200 MW, respectively—given how much was needed to meet the 2020 target. Wind was singled out ostensibly because a large number of installations already existed in regions of high potential where they could be easily repowered or expanded and biomass because of its ability to be dispatched as needed and its potential contribution to regional development. Nevertheless, the CNMC had expressed concern that the goal for biomass was too ambitious, given how little capacity (approximately 500 MW) already existed, while the limit for wind was much lower than it needed to be (CNMC 2015, 12).[3]

In keeping with the new emphasis on investment-based support, bidders would effectively offer a percentage reduction on the rate of return for the initial investment in a standard reference plant over its lifetime, and all winners would receive the rate set by the last bid needed to fill the quota. The new plants would sell their power directly to the wholesale market, and government support would be provided only if the spot price fell below the level needed to achieve the discounted rate of return. In addition, winning bidders would have to put up a guarantee of 20 euros per kilowatt (kW) of capacity and would have to complete their projects within four years (Real Decreto 947/2015; MINETUR 2015a, 2015c).

The Rajoy government had initially hoped to hold the auction before the election in December, but it was not able to finalize the details until the end of November and the auction was only able to take place the following January. Nevertheless, from the government's perspective, the auction was a success, especially with regard to wind power. Some 2500 MW in bids for wind projects were submitted, and all the winning bids—for both wind and biomass—effectively offered a 100% discount on the rate of return. Thus the government would not have to provide any support, no matter how low the wholesale market price might go (MINETUR 2016; del Río 2016, 31, 33) (Table 5.2).[4]

Nevertheless, the auction was criticized by representatives of the renewable energy industry and others on several grounds. One was the neglect of other technologies, especially solar PV. Another was the small size, especially given the number of projects in development that had been put on hold because of the moratorium. A third was the lack of any prequalification requirements. As a result, the biggest awards went to companies without strong track records in the sector, raising questions about whether or not the projects would actually be built. Finally, the government had been silent on when further auctions would be held in

Table 5.2 Auction results, 2016–2017

| Date | Capacity awarded | | | | Maximum percentage reduction in rate of return | | |
|---|---|---|---|---|---|---|---|
| | Wind | Solar PV | Biomass and other | Total | Wind | Solar PV | Biomass and other |
| Jan. 2016 | 500 | N.A. | 200 | 700 | 100 | N.A. | 100 |
| May 2017 | 2997 | 1 | 2 | 3000 | 63.4 | 51.2 | 100 |
| July 2017 | 1128 | 3909 | N.A. | 5037 | 87.1 | 69.9 | N.A. |
| Total | 4625 | 3910 | 202 | 8737 | | | |

Sources MINETUR (2016, 2017c, 2017e)

order to award the capacity deemed necessary to meet the EU 2020 target (APPA 2016; del Río 2016, 20, 31–34; 2017, 18, 20, 22, 28). One could be forgiven for agreeing with the principal renewable industry association, the Renewable Energy Business Association (*Asociación de Empresas de Energías Renovables* or APPA), that the auction was largely a public relations stunt (APPA 2015).

### *The May 2017 Auction*

Partly in response to these criticisms and in view of the continuing need to raise renewable power output by 2020, the government announced in mid-2016 that it would hold a second auction. This one would be much more substantial, covering up to 3 GW in capacity, and open to all technologies. Because of the national political crisis, during which Spain had a caretaker government for more than 10 months and was forced to hold a second round of general elections, the new Rajoy administration did not submit a draft to the CNMC until the end of December, and the details were not finalized until the following Spring. This time, the rules limited the size of the discounts on the standard rate of return that could be offered for wind (63.43%) and PV (51.22%). In the event of a tie, the award would go to the technology with the highest number of operating hours for the standard reference plant. In addition, the size of the required financial guarantee tripled to 60 euros/kW, and new plants would have to be up and running by the end of 2019 if they were to receive any support (vectorcuatro 2017a, 2017b; Real Decreto 359/2017; MINETUR 2017a, 2017b).

The second auction took place in May 2017. Once again, the auction was greatly oversubscribed, this time by approximately 300%, with bids roughly evenly divided between wind and PV. And as before, all the winning bids offered the maximum possible discount. According to one estimate, the wholesale price of electricity would have to fall below 40 euros per megawatt-hour (MWh), somewhat lower than the natural floor price at the time, before any support would be provided. Because of the provision for breaking ties, however, virtually all of the capacity (2979 out of 3000 MW) was awarded to wind projects, with a mere 1 MW going to PV and 20 MW to other technologies (MINETUR 2017c; S&P Global 2018).

Despite this success, the second auction was criticized for not being ambitious enough. It still provided for less than half the new capacity called for in the 2015 planning document. In addition, it once again triggered an outcry from the solar PV sector. The leading PV industry group, the Spanish Photovoltaic Union (*Unión Española Fotovoltaica* or UNEF), charged that PV had been discriminated against, to the benefit of wind, and the association of small PV producers, Anpier, complained that the rules favored projects with large capacities (UNEF 2017; Kenning 2017a).

### *The July 2017 Auction*

Then, just days later, the Rajoy government announced it would hold yet another auction that summer, again of up to 3 GW. What prompted such urgent action? Based on the experience of the May auction, the government realized that much more capacity was likely to be available at a deep discount in the guaranteed rate of return, and there was no time to spare if new capacity were to come online before 2020. Indeed, the third auction could award rights to more than 3 GW in new capacity in the event of tied bids, as long as any extra cost to the system was lower than a value established in advance by the government but not shared with the bidders. Because the greatest potential for low-cost capacity lay with wind and PV, however, the auction would be limited to just those technologies, but the maximum possible discount for each was raised to 87.1 and 69.9%, respectively (Real Decreto 650/2017; MINETUR 2017d).

As in May, the auction was massively oversubscribed, and the amount of capacity offered at the maximum discount exceeded 5 GW, making it the largest European renewable power auction ever in terms of capacity. This time, however, the increase in the maximum discount made PV relatively more attractive, and more than 3.9 GW of PV, versus just 1.1 GW of wind, was awarded. It also meant that the effective floor price was even lower than that of the May auction, ranging from 28 to 32 euros per MWh, depending on the project. And as with the May auction, winning bidders would have to complete their projects by the end of 2019 if they were to receive support and not to lose their guarantees (Kenning 2017b; MINETUR 2017e; S&P Global 2018).

### *Overall Assessment of the Auctions*

Taken together, the three auctions appeared to be a great success. If all the awarded capacity (8.7 GW) were built, Spain would nearly double its solar PV capacity and increase that of wind by about 20%. Altogether, according to a government estimate, the winning projects, if completed, would enable Spain to increase the renewable share of energy use to at least 19.5% by 2020, notwithstanding an expected growth in electricity consumption of nearly 1% per year (Bellini 2017a).

A number of observers, however, expressed concerns. Perhaps the greatest was that much of the awarded capacity might never materialize, even if it meant losing the required financial guarantees. The winning offers appeared too low to be profitable, the result of strategic bidding based on the assumption that the last accepted bid—which would determine the minimum guaranteed rate of return—would be higher. With such low effective floor prices, and the possibility that the government could reduce the reasonable rate of profitability yet further in the future, new plants would be almost entirely exposed to market risk. As a result, projects might find it difficult to secure financing. Second, the timely deployment of much new capacity was threatened by the potential expiration of many existing permits for grid access and connection at the end of 2018 as well as insufficient connection capacity in some regions. And many winners had little or no prior experience with big wind or solar projects (del Río 2016, 27, 34; 2018, 19, 27; Ojea 2018a, 2018b).

Each technology faced additional distinct obstacles of its own. The short-time frame was expected to be especially challenging for wind, because of potential bottlenecks in the supply chain. As for PV, the sudden growth in demand triggered a four-fold jump in land prices, potentially raising the costs of projects (del Río 2018, 29; R. Roca 2018).

Finally, even if all the capacity were to be built, there was no guarantee that it would generate the needed amounts of power. As more than one expert has noted, investment-based auctions provide no incentive to operate a plant efficiently (del Río 2017, 22).

## Market Developments

Notwithstanding these concerns, independent developments in the electricity market increased the likelihood that much of the awarded capacity would in fact be built, and may have even contributed to the willingness of so many project developers to bid at the maximum possible discount in the first place. The first of these developments was the continued and, in some cases, precipitous decline in the cost of renewable technologies. As a result, renewables were reaching the point where they could compete with traditional sources of power even without financial support, a phenomenon known as grid parity. The only question was whether the wholesale price of electricity would be high enough to cover the cost of financing any new construction.

### *Falling Costs*

Most dramatic of all was the drop in the cost of PV. Estimates vary, but by many counts, the price of components, such as silicon modules had declined by 90% over 10 years, thanks to economies of scale, and the total cost of new PV systems was down by nearly as much. By 2018, it was possible to build a new plant at an overall levelized cost of $40–60 per MWh, and even less in some parts of Spain. Put differently, PV was reaching the point where 1 MW of capacity could be constructed for just 1 million euros, and the price tag was continuing to fall (Vartiainen et al. 2020; J. Roca 2018).[5]

Although not as dramatic, the drop in the cost of wind was also significant, and it had started at a lower level than that of PV. The capital cost of wind was down some 40–70% over a decade, and that decline was paralleled by a comparable fall in operating costs (IRENA 2018, 12, 19; WindEurope 2019, 17).

### *Merchant Projects*

Not only did falling costs reduce the risks of building the capacity awarded in the auctions, but they encouraged a number of developers to announce projects that were entirely unrelated to the auctions. These so-called merchant projects would be financed entirely by revenues generated by sales to the wholesale electricity market, without any possibility of government support. This trend was reflected in a surge of applications for permits for access and connection to the electricity grid that far exceeded the volume needed to serve the 8.7 GW of capacity awarded in the auctions. In 2018 alone, the grid operator, REE, awarded 17 GW in access permits, including 11.9 GW for PV, 4.5 GW for wind, and 0.6 GW for other renewable sources (Ojea 2019a). As will be discussed below, however, not all of these permits were awarded for identified projects.

A prominent example of such merchant projects was the 500 MW Nuñez de Balboa PV plant in Extremadura. Originally conceived in 2012 just as the moratorium was going into effect, it was revived in 2017 when it received environmental approval from the government before being acquired by Iberdrola, one of Spain's major utilities (Bellini 2017b).

### Power Purchase Agreements

Nevertheless, both capacity awarded in the auctions and purely merchant projects faced challenges in securing financing, given their exposure to price risk in the wholesale market, and notwithstanding the falling costs of renewable technologies. This risk threatened only to grow, moreover, as the market penetration of low to no marginal cost renewable sources increased and drove down electricity prices.

A potential solution to this impasse was the use of power purchase agreements (PPA). Under a PPA, in its simplest form, a renewable power producer would sell its output to a consumer at a set price and for a fixed number of years. In the case of Spain, the agreed price would typically be higher than the price floor set by the auctions. The resulting guaranteed revenue stream would reduce investment risk and thus make it easier for the project developer to obtain loans and other forms of financing from third parties.[6]

A PPA would benefit the purchaser as well. It would enable the purchaser to hedge against possible upward swings in electricity costs, especially in the event of rising fossil fuel prices, and possibly even to obtain power at less than the average market price, given the declining cost of renewable installations. In addition, PPAs could help companies burnish their environmental credentials by using non-$CO_2$ producing electricity sources to power their operations.

Thanks to these potential benefits, Spain saw a steady stream of announcements for PPAs, mostly involving prospective PV plants, in the years following the auctions. The first was agreed in mid-2017. By late 2018, contracts involving 1.5 GW of capacity had been signed. And in 2019, agreements concerning another 4.4 GW of capacity were announced (UNEF 2018; Pérez Galdón 2019; El Periódico de la Energía 2020a). It is worth noting that many of the PPAs were entered into by marketers and utilities, who then sold the power they purchased from the actual producers to final consumers, with many of the rest involving electricity-intensive industries. And in at least one case, the project developer entered into a financial hedge arrangement that would lock in prices for a number of years (Ellomay Capital 2018).

In 2018, the use of PPAs received a further boost from the EU. Under the new renewables directive adopted late that year, member states were required to "assess the regulatory and administrative barriers to long-term renewables power purchase agreements, and shall remove unjustified barriers to, and facilitate the uptake of, such agreements." They were also obliged to "ensure that those agreements are not subject to disproportionate or discriminatory procedures or charges" (Directive (EU) 2018/2001).

To be sure, PPAs were not a panacea. Project developers were eager to sign contracts of up to 20 years, the amount of time typically required to amortize fully a project. Purchasers, on the other hand, tended to prefer shorter terms, given the likelihood that construction costs and electricity prices would continue to fall over time. Nevertheless, PPAs did appear to help a number of projects get built.

## Impact on Deployment

During the months following the auctions, some of the other potential obstacles to the timely deployment of the awarded capacity were reduced as well. This was especially the case after the Socialists, under Pedro Sánchez, took hold of the reins of government following the no-confidence vote against Mariano Rajoy in June 2018 and made the success of the auctions a top priority. Among other measures, the new center-left government expedited the processing of environmental impact statements, and it extended the expiration dates of existing grid connection permits to ensure that there would be adequate time to hook up the new plants (Ojea 2018c; Real Decreto-ley 15/2018). A number of regional governments, which were still responsible for approving plants of up to 50 MW, also took steps to simplify regulations, cut red tape, and reduce processing times (Ojea 2018b, 2018d; El Periódico de la Energía 2019a). And the national grid operator, REE, earmarked additional investment for the purpose of integrating new renewable capacity (Rojo Martín 2019).

Not surprisingly, increases in capacity went slowly at first, reflecting the lead times required to get new plants up and running. In 2018, there was only a small amount of growth in utility-scale wind (392 MW) and PV (26 MW), a trend that continued during the first half of 2019. But as the December 31 deadline for the capacity awarded in the 2017 auctions approached, the number of completed projects began to mount, eventually achieving a furious pace. By the end of the year, 6456 MW in new renewable capacity had been connected to the grid, an all-time annual record for Spain. Of this amount, some 5689 MW of capacity had been awarded in the 2017 auctions, just over 70% of the total of 8037 MW, and 767 MW was for merchant projects (R. Roca 2020; Monforte 2020).

The PV and wind sectors performed rather differently, however. Of the 3910 MW in PV capacity that had been awarded, 3728 MW (95%) had been completed. In contrast, only 1976 MW of new wind was in operation, less than half of the 4125 MW awarded in the 2017 auctions. One reason for the shortfall is that some developers were waiting to secure PPAs, because of the financial advantages PPAs conferred, before completing their projects, even if it meant losing the guarantees they had paid (Ojea 2019b). In addition, wind suffered from several disadvantages in comparison with PV: It had become more expensive, took longer to build, could be sited in fewer places, encountered greater difficulty in obtaining environmental permits, and faced more potential bottlenecks in the supply chain.

Nevertheless, considerable capacity was expected to be added in the following year or two. According to one early 2019 estimate, some 10 GW of solar power beyond the nearly 4 GW awarded in the auctions were in development, and another 15 GW or so of capacity was in the planning process (R. Roca 2019b; Cepeda 2018). As 2020 began, the region of Extremadura alone had more than 1300 MW of PV projects either completed but not yet in service or under construction and another 4.8 GW in the approval process (Junta de Extremadura 2020). For its part, the wind sector expected that all the 4.6 GW of capacity authorized in the three auctions would soon be completed. As of April 2019, some 11.9 GW of capacity was under development, including 6.6 GW in merchant projects, and more than 4.7 GW had construction authorization (El Periódico de la Energía 2019b; R. Roca 2019c). The autonomous community of Aragon alone had issued construction permits for more than 2.3 GW of capacity and was processing applications for another 2 GW of capacity (El Periódico de la Energía 2019c). Altogether, according to one independent analysis, another 4.3 GW of wind and 4.3 GW of solar PV would begin operation in 2020–2021 (El Periódico de la Energía 2020b).

## CONCLUSION

The deployment of new renewable capacity at the end of 2019 was truly remarkable, and more substantial than many had expected when the auctions were held just two to three years before. Most notably, the amount of utility-scale solar PV capacity in Spain nearly doubled, and many more projects were in the pipeline. Nevertheless, some questions and criticisms remained.

One important limitation was the fact that small-scale distributed generation, especially using solar PV, had been entirely neglected by the auctions. Many observers had come to believe that the promotion of distributed generation was critical to the full realization of Spain's renewable potential and that it offered important advantages over a centralized power system. In fact, a parallel set of efforts to promote so-called self-consumption had begun even before the moratorium. The efforts that were made, and the obstacles that they encountered, are the subject of the next chapter.

Other questions concerned the sustainability of the auction-based buildup, notwithstanding the amount of capacity at various stages of development. Indeed, there was a sharp drop in new deployments as the end-of-2019 deadline passed. As Pablo del Río had observed as early as 2016, the Rajoy government's approach did little to build a foundation for long-term growth in the domestic renewable power industry. Rather, it threatened to repeat the boom-bust cycle of the previous decade, and there would be little incentive to make long-term investments in the supply chain (del Río 2016, 35). Chapter 7 will discuss the actions taken by the successor governments of Pedro Sánchez to lay the groundwork for continued growth in renewable power generation.

During this period, however, we begin to witness a reduced role for the central government in promoting renewable power. The declining cost of the leading technologies, especially PV, and the approach of grid parity meant that it would be increasingly possible to finance new projects without public support. To be sure, as we shall see, the government would continue to be critical for creating a framework that fostered private investment in renewable projects, and it was indispensable to setting up the auctions of 2016 and 2017. But even then, the need to take action and the form that it took (auctions) were largely conditioned by external forces, that is, Spain's EU energy obligations and EU restrictions on the use of state aid.

## Notes

1. As before, the directive required states to guarantee grid access and priority dispatch for renewable sources (Article 16) but otherwise lacked common instruments for promoting renewables.
2. The most thorough description of the 2016 auction is del Río (2016, 11–20).
3. In 2014, the government had authorized a separate call for up to 450 MW of wind power in the Canary Islands.
4. On reasons for the low bids, see del Río (2016, 34).
5. Levelized cost of electricity (LCOE) represents the average revenue per unit of electricity generated that would be required to recover the costs of building and operating a generating plant during an assumed financial life and duty cycle. LCOE is often cited as a convenient summary measure of the overall competitiveness of different generating technologies (EIA 2021).
6. A useful description of PPAs in the Spanish context is AleaSoft Energy Forecasting (2019).

## References

AleaSoft Energy Forecasting. 2019. "Los PPAs, una oportunidad para los agentes de los mercados eléctricos." *El Periódico de la Energía* (27 March). https://elperiodicodelaenergia.com/los-ppas-una-oportunidad-para-los-agentes-de-los-mercados-electricos/.

APPA (Asociación de Empresas de Energías Renovables). 2015. "Las subastas renovables, otra muestra de la improvisación del Gobierno en materia energética" (22 April). https://appa.es/wp-content/uploads/descargas/2015/APPA_SUBASTAS_VF_mm.pdf.

———. 2016. "El resultado de la subasta eléctrica añade más incertidumbre al sector renovable" (15 Jan.). https://www.appa.es/wp-content/uploads/2018/08/20160115-APPA_ADJUDICACION_SUBASTA.pdf.

Bellini, Emiliano. 2017a. "Spanish PV Could Go Merchant but Market Must Change." *pv magazine* (3 Nov.). https://www.pv-magazine.com/2017/11/03/spanish-pv-could-go-merchant-but-market-must-change/.

———. 2017b. "Grupo Eco Gets Environmental Approval for 500 MW Solar Park in Spain." *pv magazine* (Dec. 18). https://www.pv-magazine.com/2017/12/18/grupo-eco-gets-environmental-approval-for-500-mw-solar-park-in-spain/.

Cepeda Minaya, Denisse. 2018. "El futuro solar: sin primas, con autoconsumo y PPA." *CincoDías* (30 Sept.). https://cincodias.elpais.com/cincodias/2018/09/27/companias/1538050375_356196.html.

CNMC (Comisión Nacional de los Mercados y la Competencia). 2015. "Informe sobre la propuesta de real decreto por el que se establece una convocatoria para el otorgamiento del régimen retributivo específico a nuevas instalaciones de producción de energía eléctrica a partir de biomasa en el sistema eléctrico peninsular y para instalaciones de tecnología eólica y sobre la propuesta de orden por la que se regula el procedimiento de asignación del régimen retributivo especifico en dicha convocatoria y se aprueban sus parámetros retributivos." IPN/DE/007/15 and IPN/DE/008/15 (18 June). https://www.cnmc.es/sites/default/files/1489860_8.pdf.

Congreso de los Diputados. 2010. "Informe de la subcomisión de análisis de la estrategia energética española para los próximos 25 años." *Boletín Oficial de las Cortes Generales*, no. 501 (30 Dec.).

Council of the European Union. 2007. "Brussels European Council 8/9 March 2007: Presidency Conclusions" (2 May). http://register.consilium.europa.eu/doc/srv?l=EN&f=ST%207224%202007%20REV%201.

del Río, Pablo. 2016. "Implementation of Auctions for Renewable Energy Support in Spain." Report D7.1-ES (March). http://auresproject.eu/sites/aures.eu/files/media/documents/wp7_-_case_study_report_spain_1.pdf.

———. 2017. "Assessing the Design Elements in the Spanish Renewable Electricity Auction: An International Comparison." *Working Paper 6/2017*. Real Instituto Elcano (17 April). http://www.realinstitutoelcano.org/wps/portal/rielcano_en/contenido?WCM_GLOBAL_CONTEXT=/elcano/elcano_in/zonas_in/dt6-2017-delrio-design-spanish-renewable-electricity-auction.

———. 2018. "An Analysis of the Design Elements of the Third Renewable Energy Auction in Spain." *Renewable Energy Law and Policy Review* 8, no. 3 (April): 17–30.

Directive 2009/28/EC of the European Parliament and of the Council of 23 April 2009 on the Promotion of the Use of Energy from Renewable Sources and Amending and Subsequently Repealing Directives 2001/77/EC and 2003/30/EC. *Official Journal of the European Union* (5 June 2009). https://eur-lex.europa.eu/LexUriServ/LexUriServ.do?uri=OJ:L:2009:140:0016:0062:EN:PDF.

Directive (EU) 2018/2001 of the European Parliament and of the Council of 11 December 2018 on the Promotion of the Use of Energy from Renewable Sources. *Official Journal of the European Union* (11 Dec. 2018). https://eur-lex.europa.eu/legal-content/EN/TXT/PDF/?uri=CELEX:32018L2001&from=EN.

EIA (U.S. Energy Information Administration). 2021. "Levelized Costs of New Generation Resources in the Annual Energy Outlook 2021" (Feb.). https://www.eia.gov/outlooks/aeo/pdf/electricity_generation.pdf.

Ellomay Capital. 2018. "Ellomay Capital Enters Into a Binding Term Sheet for a Power Financial Hedge for the Talasol Project" (24 Jan.). https://ellomay.com/press-releases/ellomay-capital-enters-into-a-binding-term-sheet-for-a-power-financial-hedge-for-the-talasol-project/.

El Periódico de la Energía. 2019a. "Aragón lidera èl nuevo 'boom' renovable con 79 parques eólicos y 56 fotovoltaicos" (4 March). https://elperiodicodelaenergia.com/aragon-lidera-el-sector-de-las-energias-renovables-con-79-parques-eolicos-y-56-fotovoltaicos/.

———. 2019b. "El sector eólico espera instalar más de 4.600 MW hasta 2020" (13 Feb.). https://elperiodicodelaenergia.com/el-sector-eolico-espera-instalar-mas-de-4-600-mw-hasta-2020/.

———. 2019c. "Aragón lidera el nuevo 'boom' renovable con 79 parques eólicos y 56 fotovoltaicos" (4 March). https://elperiodicodelaenergia.com/aragon-lidera-el-sector-de-las-energias-renovables-con-79-parques-eolicos-y-56-fotovoltaicos/.

———. 2020a. "España arrasa Europa en el desarrollo de proyectos renovables con PPA en 2019" (31 Jan.). https://elperiodicodelaenergia.com/espana-arrasa-europa-en-el-desarrollo-de-proyectos-renovables-con-ppa-en-2019/.

———. 2020b. "España prevé instalar otros 8.600 MW de renovables en el bienio 2020–2021" (6 Oct.). https://elperiodicodelaenergia.com/espana-preve-instalar-otros-8-600-mw-de-renovables-en-el-bienio-2020-2021/.

Energías Renovables. 2010a. "El Gobierno elimina del Pacto de Zurbano a las renovables" (9 April). https://www.energias-renovables.com/panorama/el-gobierno-elimina-del-pacto-de-zurbano.

———. 2010b. "¡Pues no se salva ni Zurbano!" (15 June). https://www.energias-renovables.com/panorama/iexcl-pues-no-se-salva-ni.

EC (European Commission). 2012. "Assessment of the 2012 National Reform Programme and Stability Programme for Spain." Commission Staff Working Document. SWD(2012) 310 final (30 May). https://eur-lex.europa.eu/legal-content/EN/TXT/PDF/?uri=CELEX:52012SC0310&from=en.

———. 2013. "European Commission Guidance for the Design of Renewables Support Schemes." Commission Staff Working Document SWD(2013) 439 final (5. Nov.). https://ec.europa.eu/energy/sites/ener/files/documents/com_2013_public_intervention_swd04_en.pdf.

———. 2014. *Guidelines on State aid for Environmental Protection and Energy 2014–2020 (2014/C 200/01)*. https://eur-lex.europa.eu/legal-content/EN/TXT/PDF/?uri=CELEX:52014XC0628(01)&from=EN.

———. 2015. "Renewable Energy Progress Report." COM(2015) 293 final (15 June). https://eur-lex.europa.eu/resource.html?uri=cellar:4f8722ce-1347-11e5-8817-01aa75ed71a1.0001.02/DOC_1&format=PDF.

IDAE (Instituto para la Diversificación y Ahorro de la Energía). 2010. *Plan de Acción Nacional de Energías Renovables de España (PANER) 2011 – 2020* (30

June). https://www.mincotur.gob.es/energia/desarrollo/EnergiaRenovable/Documents/20100630_PANER_Espanaversion_final.pdf.

———. 2011a. *Plan de Energías Renovables (PER) 2011–20* (11 Nov.). http://www.idae.es/file/9712/download?token=6MoeBdCb.

———. 2011b. *Resumen del Plan de Energías Renovables 2011–2020* (26 July). https://www.idae.es/uploads/documentos/documentos_Resumen_PER_2011-2020_26-julio-2011_58f27847.pdf.

IRENA (International Renewable Energy Association). 2018. *Renewable Power Generation Costs in 2017*. http://irena.org/-/media/Files/IRENA/Agency/Publication/2018/Jan/IRENA_2017_Power_Costs_2018.pdf.

Junta de Extremadura. 2020. "Extremadura cierra 2019 con 13 nuevas instalaciones fotovoltaicas en servicio con una potencia total de 511 megavatios" (29 Jan.). http://www.juntaex.es/comunicacion/noticia&idPub=29578#.X2ukhS9h3_R.

Kenning, Tom. 2017a. "Spain's New Renewables Tender Branded 'Unfair' by Solar Industry." *PV Tech* (11 April). https://www.pv-tech.org/news/spain-renewables-auction-progress-but-solar-industry-decries-impartiality.

———. 2017b. "Spain Awards 4GW Solar and 1GW Wind in Auction Surprise." *PV Tech* (27 July). https://www.pv-tech.org/news/spain-awards-4gw-solar-and-1gw-wind-in-auction-surprise.

Ley 24/2013, de 26 de diciembre, del Sector Eléctrico (26 Dec. 2013). https://www.boe.es/eli/es/l/2013/12/26/24.

MINETUR (Ministerio de Industria, Energía y Turismo). 2015a. "Orden IET/2212/2015, de 23 de octubre, por la que se regula el procedimiento de asignación del régimen retributivo específico en la convocatoria para nuevas instalaciones de producción de energía eléctrica a partir de biomasa situadas en el sistema eléctrico peninsular y para instalaciones de tecnología eólica, convocada al amparo del Real Decreto 947/2015, de 16 de octubre, y se aprueban sus parámetros retributivos" (24 Oct.). https://www.boe.es/boe/dias/2015/10/24/pdfs/BOE-A-2015-11432.pdf.

———. 2015b. *Planificatión Energética: Plan de Desarrollo de la Red de Transporte de Energía 2015–2020*. https://www.mincotur.gob.es/energia/planificacion/Planificacionelectricidadygas/desarrollo2015-2020/Documents/Planificaci%C3%B3n%202015_2020%20%202016_11_28%20VPublicaci%C3%B3n.pdf.

———. 2015c. "Resolución de 30 de noviembre de 2015, de la Secretaría de Estado de Energía, por la que se convoca la subasta para la asignación del régimen retributivo específico a nuevas instalaciones de producción de energía eléctrica a partir de biomasa situadas en el sistema eléctrico peninsular y a instalaciones de tecnología eólica, y se establecen el procedimiento y las reglas de la misma, al amparo de lo dispuesto en el Real Decreto 947/2015, de 16 de octubre, y en la Orden IET/2212/2015, de 23 de octubre" (3 Dec.). https://www.boe.es/diario_boe/txt.php?id=BOE-A-2015-13124.

———. 2016. "Resolución de 18 de enero de 2016, de la Dirección General de Política Energética y Minas, por la que se resuelve la subasta para la asignación del régimen retributivo específico a nuevas instalaciones de producción de energía eléctrica a partir de biomasa en el sistema eléctrico peninsular y para instalaciones de tecnología eólica, al amparo de lo dispuesto en el Real Decreto 947/2015, de 16 de octubre" (21 Jan.), https://www.boe.es/diario_boe/txt.php?id=BOE-A-2016-552.

———. 2017a. "Orden ETU/315/2017, de 6 de abril, por la que se regula el procedimiento de asignación del régimen retributivo específico en la convocatoria para nuevas instalaciones de producción de energía eléctrica a partir de fuentes de energía renovables, convocada al amparo del Real Decreto 359/2017, de 31 de marzo, y se aprueban sus parámetros retributivos" (8 April). https://www.boe.es/eli/es/o/2017/04/06/etu315.

———. 2017b. "Resolución de 10 de abril de 2017, de la Secretaría de Estado de Energía por la que se convoca subasta para la asignación del régimen retributive específico a nuevas instalaciones de producción de energía eléctrica a partirde fuentes de energía renovables, al amparo de lo dispuesto en la Orden ETU/315/2017, de 6 de abril" (12 April). https://www.boe.es/boe/dias/2017/04/12/pdfs/BOE-A-2017-4094.pdf.

———. 2017c. "Resolución de 19 de mayo de 2017, de la Dirección General de Política Energética y Minas, por la que se resuelve el procedimiento de subasta para la asignación del régimen retributivo específico al amparo de lo dispuesto en el Real Decreto 359/2017, de 31 de marzo, y en la Orden ETU/315/2017, de 6 de abril" (26 May). https://www.boe.es/diario_boe/txt.php?id=BOE-A-2017-5848.

———. 2017d. "Orden ETU/615/2017, de 27 de junio, por la que se determina el procedimiento de asignación del régimen retributivo específico, los parámetros retributivos correspondientes, y demás aspectos que serán de aplicación para el cupo de 3.000 MW de potencia instalada, convocado al amparo del Real Decreto 650/2017, de 16 de junio" (28 June). https://www.boe.es/boe/dias/2017/06/28/pdfs/BOE-A-2017-7389.pdf.

———. 2017e. "Resolución de 27 de julio de 2017, de la Dirección General de Política Energética y Minas, por la que se resuelve el procedimiento de subasta para la asignación del régimen retributivo específico al amparo de lo dispuesto en el Real Decreto 650/2017, de 16 de junio, y en la Orden ETU/615/2017, de 27 de junio" (28 July). https://www.boe.es/diario_boe/txt.php?id=BOE-A-2017-8997.

Monforte, Carmen. 2020. "Las renovables baten récord con 6.500 MW nuevos en 2019." *CincoDías* (24 Jan.). https://cincodias.elpais.com/cincodias/2020/01/23/companias/1579810297_844217.html.

Ojea, Laura. 2018a. "Las renovables temen que el RD de acceso y conexión no llegue a tiempo y pierdan sus permisos y avales." *El Periódico de la Energía* (4 July). https://elperiodicodelaenergia.com/las-renovables-temen-que-el-rd-de-acceso-y-conexion-no-llegue-a-tiempo-y-pierdan-sus-permisos-y-avales/.

———. 2018b. "Todos los proyectos de renovables ganadores de las subastas estarán en funcionamiento el 1 de enero de 2020." *El Periódico de la Energía* (24 Sept.). https://elperiodicodelaenergia.com/todos-los-proyectos-de-renovables-ganadores-de-las-subastas-estaran-en-funcionamiento-el-1-de-enero-de-2020/.

———. 2018c. "El Gobierno agilizará los trámites de impacto ambiental de las nuevas plantas de renovables para llegar al 20% en 2020." *El Periódico de la Energía* (22 June). Retrieved from https://elperiodicodelaenergia.com/el-gobierno-agilizara-los-tramites-de-impacto-ambiental-de-las-nuevas-plantas-de-renovables-para-llegar-al-20-en-2020/.

———. 2018d. "La Comunidad Valenciana, a punto de perder el tren del 'boom' fotovoltaico en España." *El Periódico de la Energía* (23 July). https://elperiodicodelaenergia.com/la-comunidad-valenciana-a-punto-de-perder-el-tren-del-boom-fotovoltaico-en-espana/.

———. 2019a. "Locura renovable en España: quieren instalarse más de 87 GW, la mitad ya con el permiso de acceso concedido por REE." *El Periódico de la Energía* (31 Jan.). https://elperiodicodelaenergia.com/locura-renovable-en-espana-quieren-instalarse-mas-de-87-gw-la-mitad-ya-con-el-permiso-de-acceso-concedido-por-ree/.

———. 2019b. "Los proyectos renovables ganadores de las subastas prefieren perder los avales a cambio de firmar un PPA." *El Periódico de la Energía* (26 Sept.). Retrieved from https://elperiodicodelaenergia.com/los-proyectos-renovables-ganadores-de-las-subastas-prefieren-perder-los-avales-a-cambio-de-firmar-un-ppa/.

Partido Popular. 2015. *Seguir Avanzando, 2016–2020: Programa Electoral Para las Elecciones Generales de 2015.* http://www.pp.es/sites/default/files/documentos/programa2015.pdf.

Pérez Galdón, Beatriz. 2019. "Don Rodrigo inicia la nueva era de plantas solares sin subvención." *CincoDías* (9 Nov.). https://cincodias.elpais.com/cincodias/2018/11/09/companias/1541758756_915913.html.

Pérez Rodriguez, Daniel. 2015. "Resumen Sésion informative subasta biomasa y eólica." https://www.holtropblog.com/es/index.php/44-renovables/738-resumen-sesion-informativa-subasta-biomasa-y-eolica.

Planelles, Manuel. 2017. "Spain's Use of Renewable Energy Sources Stagnates." *El País* (15 March). https://english.elpais.com/elpais/2017/03/15/inenglish/1489563590_304234.html.

Real Decreto 413/2014, de 6 de junio, por el que se regula la actividad de producción de energía eléctrica a partir de fuentes de energía renovables, cogeneración y residuos (10 June 2014). https://www.boe.es/eli/es/rd/2014/06/06/413.

Real Decreto 947/2015, de 16 de octubre, por el que se establece una convocatoria para el otorgamiento del régimen retributivo específico a nuevas instalaciones de producción de energía eléctrica a partir de biomasa en el sistema eléctrico peninsular y para instalaciones de tecnología eólica (16 Oct. 2015). https://www.boe.es/eli/es/rd/2015/10/16/947.

Real Decreto 359/2017, de 31 de marzo, por el que se establece una convocatoria para el otorgamiento del régimen retributivo específico a nuevas instalaciones de producción de energía eléctrica a partir de fuentes de energía renovables en el sistema eléctrico peninsular (1 April 2017). https://boe.es/buscar/doc.php?id=BOE-A-2017-3639.

Real Decreto 650/2017, de 16 de junio, por el que se establece un cupo de 3.000 MW de potencia instalada, de nuevas instalaciones de producción de energía eléctrica a partir de fuentes de energía renovables en el sistema eléctrico peninsular, al que se podrá otorgar el régimen retributivo específico (17 June 2017). https://www.boe.es/diario_boe/txt.php?id=BOE-A-2017-6940.

Real Decreto-ley 1/2012, de 27 de enero, por el que se procede a la suspensión de los procedimientos de preasignación de retribución y a la supresión de los incentivos económicos para nuevas instalaciones de producción de energía eléctrica a partir de cogeneración, fuentes de energía renovables y residuos (28 Jan. 2012). https://www.boe.es/eli/es/rdl/2012/01/27/1.

Roca, José. 2018. "Ya es más rentable construir nuevos proyectos de renovables que mantener las plantas de generación convencionales." *El Periódico de la Energía* (12 Nov.). https://elperiodicodelaenergia.com/ya-es-mas-rentable-construir-nuevos-proyectos-de-renovables-que-mantener-las-plantas-de-generacion-convencionales/.

Roca, Ramón. 2018. "La especulación regresa a las renovables: los propietarios de suelo cuadruplican el precio de sus terrenos en un año." *El Periódico de la Energía* (4 July). https://elperiodicodelaenergia.com/la-especulacion-regresa-a-las-renovables-los-propietarios-de-suelo-cuadruplican-el-precio-de-sus-terrenos-en-un-ano/.

———. 2019a. "Récord de instalación de renovables en España en 2019: se conectaron 6.456 MW de los que 5.689 MW pertenecen a las subastas." *El Periódico de la Energía* (23 Jan.). https://elperiodicodelaenergia.com/record-de-instalacion-de-renovables-en-espana-en-2019-se-conectaron-6-456-mw-de-los-que-5-689-mw-pertenecen-a-las-subastas/.

———. 2019b. "El renacimiento solar de Europa de la mano de España: se instalarán 124 GW en los próximos cinco años." *El Periódico de la Energía* (29 April). https://elperiodicodelaenergia.com/el-renacimiento-solar-de-europa-de-la-mano-de-espana-se-instalaran-124-gw-en-los-proximos-cinco-anos/.

———. 2019c. "Fiebre por la eólica en España: hay solicitados cerca de 25 GW y casi la mitad están ya bastante avanzados." *El Periódico de la Energía* (24 April). https://elperiodicodelaenergia.com/fiebre-por-la-eolica-en-espana-hay-solicitados-cerca-de-25-gw-y-casi-la-mitad-estan-ya-bastante-avanzados/.

———. 2020. "Récord de instalación de renovables en España en 2019: se conectaron 6.456 MW de los que 5.689 MW pertenecen a las subastas." *El Periódico de la Energía* (23 Jan.). https://elperiodicodelaenergia.com/record-de-instalacion-de-renovables-en-espana-en-2019-se-conectaron-6-456-mw-de-los-que-5-689-mw-pertenecen-a-las-subastas/.

Rojo Martín, José. 2019. "Spanish Grid Set for Funding Boom to Accommodate Renewables." *PV Tech* (22 March). https://www.pv-tech.org/news/spanish-grid-set-for-funding-boom-to-accommodate-renewables.

S&P Global. 2018. "The End to Subsidies: The Beginning of a New Era for Spanish Renewables" (7 Feb. 2018). https://www.capitaliq.com/CIQDotNet/CreditResearch/RenderArticle.aspx?articleId=1990704&SctArtId=448388&from=CM&nsl_code=LIME&sourceObjectId=10410209&sourceRevId=3&fee_ind=N&exp_date=20280213-22:39:51.

UNEF (Unión Española Fotovoltaica). 2017. "El resultado de la subasta demuestra que la fotovoltaica ha sido discriminada" (17 May). https://unef.es/2017/05/el-resultado-de-la-subasta-demuestra-que-la-fotovoltaica-ha-sido-discriminada/.

———. 2018. "En España ya hay más de 1500 MW comercializados en contratos PPAs" (10 Oct.). https://unef.es/2018/10/en-espana-ya-hay-mas-de-1500-mw-comercializados-en-contratos-ppas/.

Vartiainen, Eero, Gaëtan Masson, Christian Breyer, David Moser, and Eduardo Román Medina. 2020. "Impact of Weighted Average Cost of Capital, Capital Expenditure, and Other Parameters on Future Utility-Scale PV Levelised Cost of Electricity." *Progress in Photovoltaics: Research and Applications* 28: 439–453. https://onlinelibrary.wiley.com/doi/epdf/10.1002/pip.3189.

vectorcuatro. 2017a. "New Renewable Energy Auction to Be Held in Spain on 17 May 2017: Are You Ready?" (19 April). http://www.vectorcuatrogroup.com/en/new-renewable-energy-auction-to-be-held-in-spain-on-17-may-2017-are-you-ready/.

———. 2017b. "Spain to Host New Renewable Energy Auctions: Restart or Hurdles?" (18 Jan.). http://www.vectorcuatrogroup.com/en/spain-to-host-new-renewable-energy-auctions-restart-or-hurdles/.

WindEurope. 2019. *Financing and Investment Trends: The European Wind Industry in 2018.* https://windeurope.org/wp-content/uploads/files/about-wind/reports/Financing-and-Investment-Trends-2018.pdf.

# CHAPTER 6

# The Battle Over Self-Consumption

## INTRODUCTION

In contrast to the considerable cross-party consensus Spain has seen for promoting—and occasionally limiting support for—utility-scale renewable power, the country has experienced a high degree of political conflict over small-scale electricity generation primarily for immediate consumption by the producer (or prosumer) himself, a practice commonly known as "self-consumption" (*autoconsumo*). The systematic development of explicit regulation of self-consumption was almost an afterthought, not beginning until efforts to limit the consequences of the initial boom in renewable capacity were already well underway. The process soon elicited an unprecedented level of political discord, however, with the People's Party pitted against virtually all the other national parties over the degree to which self-consumption should be encouraged. As a result, many of the obstacles to widespread deployment were not lifted until the conservative government of Mariano Rajoy was deposed and the much more favorably disposed Socialist government of Pedro Sánchez took office.

The degree of conflict may seem particularly surprising in view of the relatively low stakes involved. Many experts expected self-consumed electricity never to amount to more than a small share of renewable power production. Nevertheless, support for self-consumption became bound up with the issues of whether and how to support renewable power more generally and how to deal with the financial fallout of the boom in solar photovoltaic (PV) systems.

This chapter details the battle over self-consumption from its origins in the early 2010s until its seeming resolution at the end of the decade. It begins with the initial, incomplete attempts to develop the needed regulations and then describes the obstacles erected by the Rajoy government after it took office in late 2011. A further section covers the repeated efforts made by other parties to remove those obstacles and the success of the Rajoy government at resisting those efforts until it was cast out of office. The last sections detail the steps taken by the subsequent Sánchez government to remove the barriers and some of the additional issues that needed to be resolved before the full potential of self-consumption could begin to be realized.

## Why Self-Consumption?

Before examining the successive efforts that have been made to shape policy and regulation toward self-consumption, it is necessary to clarify why this has been regarded as such an important issue. Self-consumption has been defined as the consumption of electricity coming from generation facilities connected inside a consumer's internal network or through a direct line connecting the generator to a consumer (UNEF 2014, 87). In brief, it is the production of electricity on the customer's side of the electric meter (Mir-Artigues et al. 2018). All of the power so produced need not be consumed instantaneously by the consumer, however. In theory, any excess production can be fed into the distribution grid for use by other consumers and compensated for in various ways. Finally, self-consumption has traditionally been associated with renewable sources of power. Although this need not be the case, it has been true of Spain, especially with regard to solar PV.

The production of electricity from renewable sources for self-consumption has been thought to offer many, if not all, of the usual benefits of renewable power generation. Like the latter, it can reduce carbon emissions, increase energy independence, and strengthen a country's balance of payments by lowering the use of imported fossil fuels and the financial flows associated with them. In addition, it can create jobs and stimulate a country's renewable power industry while exerting downward pressure on the wholesale price of electricity.

At the same time, self-consumption is thought to offer a number of advantages over the traditional model of centralized generation. By co-locating production and consumption, it enhances security of supply and reduces the losses—as much as 10–13%—associated with the transportation and distribution of electricity, possibly even reducing the need to invest in the grid. Prosumers can stabilize their electricity costs by reducing their exposure to fluctuations in market prices. The construction and maintenance of such highly distributed power generation facilities can provide local jobs and generate wealth in less developed areas while minimizing the environmental impact of installations and increasing social acceptance.

At a more fundamental level, some view self-consumption as a basic right that is central to the development of a more democratic and sustainable energy model. It puts prosumer citizens at the center of decision-making about the production, distribution, and consumption of energy, enabling them to be protagonists in the electricity system (EnerAgen, n.d.; Alianza por el Autoconsumo 2017).

This is not to say that self-consumption has no disadvantages in comparison with traditional, utility-scale renewable power generation. Above all, such distributed forms of generation are generally more expensive than utility-scale plants, notwithstanding falling costs, given economies of scale. Depending on how accurately consumption and production can be forecast, moreover, self-consumption can increase the difficulty of managing the electrical system in order to ensure that supply and demand remain in balance.

## Initial Efforts to Promote and Regulate Self-Consumption

The 1980s and 1990s saw the first efforts to promote the closely related activities of self-generation and self-production, but little came of these measures and they were eventually superseded by the broader effort to promote renewable power embodied in the special regime. Sustained and systematic interest in promoting self-consumption did not materialize until the end of the decade of the 2000s, and the Socialist government of José Luis Rodríguez Zapatero made the first attempt to craft the necessary enabling regulation shortly before leaving office at the end of 2011. Initially, the incoming Rajoy government showed some interest in pursuing this work, but as we shall see in the next section, this attitude soon changed.

*Precursors*

Interest in self-consumption dates back to the earliest energy legislation under the constitution of 1978. The Energy Conservation Law of 1980 devoted an entire chapter to the promotion of what was then called self-generation (*autogeneración*) of electricity and small-scale (up to 5 megawatts (MW)) hydroelectric power. It established the rights of self-generators to connect to the grid and to transfer any excess electricity to the distribution company at a regulated price (Ley 82/1980). This law was followed by detailed regulations for the promotion of self-generation of electric power, and by the end of the decade, self-generators, including both renewable power and co-generation, were providing nearly 5% of the electricity supply. In view of this progress, the 1991–2000 National Energy Plan called for increasing the contribution of self-generation to 10% by 2000 (BOCG 1991).

Within a few years, however, interest in self-generation waned as large-scale renewable power production emerged as a distinct possibility. The 1994 regulation that established the first special regime for renewable and co-generation facilities of up to 100 MW placed much less emphasis on self-generation, which had increasingly involved the use of thermal power plants with ever greater capacity (Real Decreto 2366/1994). And although the 1997 law on the electrical sector, which introduced competition in power generation and sales, recognized what were now called self-producers (autoproductores), they would be treated no differently than would other installations in the ordinary and special regimes as a function of their size and means of power generation. Moreover, only self-producers that used co-generation or other forms of production associated with non-electrical activities, such as heat, were included in the special regime (Ley 54/1997; Mir-Artigues 2013, 668).[1]

As a result of these changes, little came of the attention initially devoted to self-generation in the 1980s and early 1990s. The renewable energy plans of 1999 and 2005 made little mention of—and set no targets for—self-production or self-consumption (IDAE 1999, 2005). Instead, for the next decade or so, Spanish efforts to promote renewable power focused on ground-mounted, grid-connected, utility-scale facilities, culminating in the boom described in Chapter 4.

## Renewed Interest

It was not until the end of the 2000s that self-consumption began to receive serious consideration in national energy planning and regulation. One reason for this development was political. In 2009, the European Union (EU) updated its directive on renewable energy. The EU now advocated decentralized energy production, noting its many benefits. To this end, the directive called for simplified and less burdensome authorization procedures for smaller projects and decentralized renewable energy production (Directive 2009/28/EC, 16, 21, 33). A second factor was economic: the declining costs of renewable power generating technologies, especially solar photovoltaic (PV), made it now possible to anticipate the eventual advent of grid parity, when the cost of producing one's own electricity would be no greater than that of buying it from the big utilities.

These developments, plus concerns about the impact on the renewable power industry of growing restrictions on government support, were reflected in an ambitious proposal put forward in late 2009 by the Photovoltaic Industry Association (*Asociación de la Industria Fotovoltaica* or ASIF). Based on an analysis by the consulting firm KMPG, the proposal assumed that grid parity would arrive in 2016. Until then, self-consumed electricity would receive an incentive equal to the difference between the value of the electricity saved and the remuneration the producer would otherwise obtain under the existing support system (ASIF 2009; see also Mir-Artigues 2012, 407–410; Jacobs 2012, 146).

Although nothing came of this particular proposal, self-consumption soon began to receive attention in Spain's energy plans. The National Renewable Energy Action Plan (*Plan de Acción Nacional de Energías Renovables de España* or PANER) for 2011–2020, issued in June 2010, called for favoring renewable power generation intended for self-consumption by providing compensation for excess production. The plan expected, as of 2015, a growing penetration in particular of PV self-consumption systems as cost parity was achieved (IDAE 2010, 50, 151).

The subsequent Renewable Energy Plan (*Plan de Energías Renovables* or PER) for 2011–2020, published in November 2011, developed these ideas in greater detail. Overall, it proposed increasing the amount of distributed generation by promoting self-consumption. The principal issue to be resolved was what to do with any electricity produced that could not be used instantaneously by the prosumer. Building distributed storage systems would be prohibitively expensive. Instead, the plan proposed using the pre-existing electrical system to absorb excess production under a compensatory arrangement known as net metering (*balance neto*). Self-producers would effectively receive credit for any excess electricity fed into the grid. This credit could then be drawn upon in the form of electricity from the grid at times when need exceeded self-production. To allow net metering, however, new legislation and regulations would be required. The plan also proposed a modest financing scheme for self-consumption installations of less than 10 kilowatts (kW) (IDAE 2011, xlii, 326, 386, 388, 492–493, 504, 538–539, 557–558; see also Mir-Artigues 2013, 669, and Mir-Artigues et al. 2018).

### *Attempts by the Socialists to Regulate Self-Consumption*

In fact, the Zapatero government, in power since 2004, was already at work developing regulations that would facilitate self-consumption as part of a broader effort to promote distributed generation. In mid-2010, it issued draft regulations for simplifying the process of establishing grid connections for facilities of less than 100 kW. After two rounds of review by the energy regulator, the National Energy Commission (*Comisión Nacional de Energía* or CNE), the government formally adopted a revised set of regulations (Real Decreto 1699/2011) in November 2011, just two days before that year's general election. It covered self-consumption generation facilities up to 100 kW or the capacity of the consumer's contracted connection to the distribution grid, whichever was smallest. But *Real Decreto* 1699/2011 left comprehensive regulation of self-consumption, including how to treat excess production, for a second round of measures that would be proposed by the Ministry of Industry within four months of its going into force (Real Decreto 1699/2011; Mir-Artigues 2013, 669; UNEF 2013a, 23–24).

Reflecting its keen interest in the subject, the Zapatero government issued a draft of the needed regulation that very day. Consistent with the renewable energy plan published that same month, the draft proposed that excess electricity fed into the local distribution grid could be banked for compensation at a later time under a net metering arrangement. In any case, however, work on the regulation halted temporarily as the People's Party took over the reins of government the following month after its triumph in the November national election (Mir-Artigues 2013, 668–669; Mir-Artigues et al. 2018).

### Conservatives: Initial Hints of Support

Initially, the new Rajoy government seemed favorably disposed to pick up where the Socialists had left off. One of its first measures, *Real Decreto-ley* 1/2012, which halted support payments for new renewable installations under the special regime, nevertheless acknowledged that the distributed generation model was becoming more and more important and noted that regulation was still being developed for the net metering procedure, calling it a real alternative for the promotion of small installations. And a subsequent measure (Real Decreto-ley 13/2012) modified the legal definitions of producer and consumer to allow the government to establish modalities of supply in order to promote individual production of electricity for consumption in the same location. Meanwhile, the CNE went forward with the process of reviewing the November 2011 draft regulation prepared by the Zapatero government, issuing a report the following March that raised questions about net metering and called for a comprehensive review of the access tariff paid by consumers (CNE 2012).

## THE RAJOY GOVERNMENT'S COUNTERREFORMATION

Over the following year, however, the stance of the new government shifted in a decidedly less favorable direction. Under pressure from the major utilities, and reflecting its more conservative disposition, the Rajoy government issued in mid-2013 a revised draft regulation that bore little resemblance to the original version. Perhaps most importantly, it contained no provision for compensating excess production fed into the grid and it established a special tax on the self-consumed portion that was widely viewed as making self-consumption unaffordable. As a result, the new draft triggered an outcry by the many supporters of self-consumption, prompting the government to shelve the proposal for two entire years. Then, in 2015, as the next general election loomed and the People's Party's hold on power appeared increasingly tenuous, the government issued a second draft that was hardly any different from the previous one and, after subjecting it to the required review process, approved it with only minor changes.

## Growing Opposition

The first resistance to self-consumption appeared among the major electrical utilities. In April 2012, one of them, Iberdrola, released an analysis arguing that, as currently proposed, self-consumption would be subsidized by other consumers, who paid a variable charge on the electricity they purchased in order to help cover the costs of the electricity system, and that this cross-subsidy would promote uncontrolled development, as had recently occurred with utility-scale PV installations. The analysis concluded that self-consumers should pay a backup charge in order to cover their share of the system costs (Sáenz de Miera 2012). Some observers noted that the utilities had their own interests at stake as well. Over the previous decade, Iberdrola and others had invested heavily in natural gas-fired power plants, which were already operating at a very low capacity, and they stood to lose even more market share if self-consumption were allowed to flourish (Barrero 2017a).

## The Rajoy Government's Initial Proposal

The concerns of the utilities were eventually reflected in the measure developed by the Rajoy government to regulate self-consumption. The first step was contained in an urgent measure adopted to ensure the financial stability of the electricity system in mid-2013, *Real Decreto-ley* 9/2013, which established a mandatory registry for self-consumers so that their activities could be monitored. This step was immediately followed by a detailed regulatory proposal on self-consumption as part of a comprehensive package of reforms of the electricity sector. The proposal bore almost no resemblance to its predecessor and, in fact, seemed expressly designed to discourage self-consumption. Most notably, as had been advocated by the big utilities, it imposed a backup charge (*peaje de respaldo*) on self-consumed electricity, on top of the fixed access charge based on the customer's maximum load and which would apply retroactively to existing installations. In addition, any excess electricity fed into the grid would not be compensated. And any facilities that failed to register would be subject to potentially massive fines (MINETUR 2013).

The government justified these unfavorable terms on several grounds. First, it argued, Spain already possessed considerable generating capacity. Indeed, total capacity exceeded peak demand by more than 60% (Roca 2013). Second, self-consumption would shift the costs of maintaining the electrical system to customers who did not self-generate, including some who could least afford to pay more. Third, self-consumption would not allow for any reduction in the capacity of the power system even as it reduced the volume of electricity that passed through it. As a matter of principle, then, self-consumers should contribute on an equal basis to the general costs of the system, which had nothing to do with whether someone was self-consuming or not. Making others pay all these costs would be unjust.

Not surprisingly, the proposal was roundly criticized by the renewable power industry, led by the Spanish Photovoltaic Union (*Unión Española Fotovoltaica* or UNEF), which was counting on investment in self-consumption to help it survive the moratorium on new support for traditional renewable power plants. The critics quickly dubbed the new charge the "sun tax" (*impuesto al sol*) and noted that it would in fact be greater than the variable access fees paid by traditional consumers, making self-generated electricity more expensive than purchases from conventional utilities and thereby rendering investment in self-consumption economically prohibitive. They also complained that co-generators would be exempted from paying the backup charge until 2020. In its own analysis, the CNE found the proposal to be "discriminatory" in comparison with other measures consumers could take to reduce their purchases from the grid and called for eliminating the backup charge (UNEF 2013b, 2014, 37, 92–95; CNE 2013a; Roca 2013).

### *Laying the Groundwork*

In the face of this criticism, the Rajoy government put further development of the regulations on hold while laying the groundwork for future action with its comprehensive overhaul of the electricity sector. The new law, *Ley 24/2013*, adopted at the end of 2013, noted that its purpose was "to guarantee an orderly development of [self-consumption], compatible with the need to guarantee the technical and economic sustainability of the electrical system as a whole." In an entire article devoted to self-consumption (Art. 9), it now distinguished between three modalities: supply with self-consumption, for those who generated electricity only for their own use; production with self-consumption, for those who registered their generating facilities as regular power plants; and production with self-consumption by a consumer connected by a direct line to a generating facility that was registered as a power plant. Elaborating on the preamble, the law affirmed that self-consumers would be "obliged to pay the same access tolls to the networks, charges associated with the system costs, and costs for the provision of system backup services" corresponding to other consumers, and it confirmed their obligation to register in the administrative registry created in July. Failure to register, as well as the use of unauthorized compensation measures or the violation of technical requirements resulting in disruptions to the grid, would be considered very serious infractions (*infracciones muy graves*), subject to fines of at least 6 million euros and possibly as high as 60 million euros. The government would establish the administrative, technical, and economic conditions for self-consumption facilities, including the sale of excess electricity to the network, at a later date.

As with the self-consumption regulation, the relevant provisions of the first draft of the new law were criticized by the CNE and especially by the National Competition Commission (*Comisión Nacional de la Competencia* or CNC). The latter found that the draft would allow the introduction of unnecessary and disproportionate obstacles to the development of self-consumption, which it found to be a healthy source of competition that could reduce the cost of electricity. The government, however, largely ignored both bodies' recommendations (CNE 2013b; CNC 2013; Barrero 2013a; Transparency International España 2014).

### *Second Draft of the Regulations (5 June 2015)*

Despite the promise of detailed regulation contained in the new electricity law, the government did not present a revision of the July 2013 draft until June 2015, as the next general election loomed. The revised document did little to appease the critics. On the contrary, all the elements that had previously raised concern remained: the retroactive application to pre-existing installations, the lack of compensation for excess electricity fed into the grid, fines of up to 60 million euros for failing to register a facility, and the controversial charge on self-consumed electricity, although the latter had been modified to exclude the cost of maintaining the transportation and distribution grids but to include balancing services necessitated by the presence of self-consumption as well as other non-grid-related system costs. In addition, the new draft introduced more complicated administrative procedures and added taxes on storage facilities, such as batteries, connected to the consumer's internal network. Meanwhile, conventional fossil fuel and nuclear power plants, which consumed approximately 8% of their own production, remained, like co-generation facilities, exempt from any of these charges (Kenning 2015a; El País 2015; Castillo et al. 2015; Greenpeace 2015; Europa Press 2015; Energías Renovables 2015a).

The government was also criticized for the process by which the draft had been developed. It had made little effort to address the concerns previously expressed by key stakeholders. It had ignored the relevant EU directives promoting distributed generation and the corresponding efforts by other EU states. And after sitting on the project for nearly two years, it was now attempting to obtain quick approval, allowing only 15 working days for comments. In less than a week, petitions circulated by critics gathered 180,000 signatures (Castillo et al.; Kenning 2015b).

The obligatory analysis prepared by the National Commission for Markets and Competition (*Comisión Nacional de los Mercados y la Competencia* or CNMC), which had subsumed the CNE and the CNC in early 2013, also found the draft to be wanting in a number of respects. It lacked a methodology for calculating the charges. It left in limbo installations of more than 100 kW. It disincentivized storage. And given the potential contribution of self-consumption to achieving Spain's environmental goals, the draft inappropriately included the costs of supporting other renewables (CNMC 2015).

The government quickly revised the proposal and submitted it for final approval in late July. The new draft helped small installations of no more than 10 kW, exempting them from the variable tax on self-consumption and simplifying their administrative procedures. It also made concessions on the use of storage facilities. But overall, the proposal continued to raise substantial barriers to the development of self-consumption and thus did little to placate the critics (El Periódico de la Energía 2015; Roca 2015; Kenning 2015c).

A final step in the adoption of the new regulations was a non-binding review by the State Council (*Consejo de Estado*), the highest consultative organ in the government. In its September report, the council proposed multiple revisions, including the elimination of the sun tax, financial compensation for excess electricity fed into the grid, and support for storage systems. Once more, the government ignored this external advice and approved the regulations without modification on October 9 (Castillo et al. 2015).

## Mission Accomplished: Real Decreto 900/2015

In announcing the new regulations, the government reiterated the principle that all consumers must contribute to the general costs of maintaining the electrical system. In its view, to exempt self-consumers from that obligation would be unjust and regressive. Moreover, it argued, distributed generation did not reduce the costs of maintaining the transmission and distribution networks and other shared costs of the system, and in some cases, it could require additional investments (Presidencia 2015; MINCOTUR 2015). Above all, the conservative government continued to prioritize reducing the tariff deficit over other possible policy goals, including the promotion of renewables.

The final version, published as *Real Decreto* 900/2015, consisted of a sprawling 44 pages. The regulations addressed all forms of self-consumption, even those without the technical possibility of exporting excess production to the grid. The only exemptions to charges were for isolated installations not in any way connected to the grid and emergency backup power systems. With minor exceptions, all others would be subject to three charges: a grid access fee based on the self-consumer's contractual capacity with the distribution network, a variable charge for electricity imported from the grid, and the variable charge on self-consumed electricity. They would also have to register in a special administrative registry for self-consumption (Title 6), as previously called for, or be subject to a fine of up to 60 million euros (Real Decreto 900/2015).

Of those subject to the various charges, the regulations simplified the three modalities identified in the law down to just two. Type 1 corresponded to supply with self-production while Type 2 combined the two modalities of production with self-consumption. In the case of Type 1, installed generation capacity was limited to the self-consumer's contractual capacity, and in any case no more than 100 kW. The self-consumer would receive no remuneration for any excess power fed to the grid and would have to install two separate meters. As a transitional measure, self-consumers with a capacity of no more than 10 kW would be exempt from some charges. Type 2 self-consumers could generate more than 100 kW and sell exported production on the wholesale market at the spot price. Like other generators, however, they would have to register as electric production installations and be subject to a generation tax of 7% and a grid connection fee (0.5 euros per megawatt-hour) as well as the charges paid by all self-consumers. The regulation also prohibited the sharing of a single generating facility by multiple consumers.

Critics of the regulation universally expected that it would have a strongly dissuasive effect on investment in self-consumption. Indeed, many of them viewed it as a deliberate attempt to do so. As far as the outcome was concerned, they were not mistaken. Self-consumption grew only very slowly during the years of the Rajoy government, despite a continuing decline in the cost of PV technology in particular, and what investment did occur largely took the form of isolated installations in the agricultural sector (Mir-Artigues et al. 2018; UNEF 2016, 69–70).[2]

## POLITICAL COUNTERPRESSURE: EFFORTS TO REVERSE *REAL DECRETO* 900/2015

The adoption of *Real Decreto* 900/2015 did not end the debate, however. On the contrary, it was followed by several years of repeated attempts to reverse it or at least to limit its impact. Many of these took place in the national political arena, where virtually all the other political parties banded together to propose more favorable regulation. In addition, efforts to promote self-consumption went forward at the regional level, while industry associations and other interest groups became increasingly active and well coordinated. And relief was pursued in the judicial arena, although with mixed results.

### National Political Efforts

Even before the ink was dry on *Real Decreto* 900/2015, the government's position on self-consumption encountered widespread political opposition. Indeed, concerted political action had begun as early as July 2015, when the government's latest draft regulation was being discussed. 18 parties, including all the major challengers, signed a manifesto calling on the government to withdraw the proposal and pledging to revoke it and support regulation favorable to the development of self-consumption should they participate in the next government, a pledge that was repeated just a month before the December election (Kenning 2015c; Energías Renovables 2015b).

Then, in February 2016, as parties maneuvered to form a new government, a majority (227) of the newly elected deputies, again representing all the opposition parties, signed an agreement committing themselves to amending the Electricity Sector Law in order to promote self-consumption within 100 days of forming a government. In particular, they would provide for compensation for excess production fed into the grid, simplify administrative procedures, and recognize the rights to self-consumption without any special charges and to sharing generating facilities (Kenning 2016; Acuerdo 2015).

When it became clear that a second election would be needed, the Citizens party (*Ciudadanos*), which had only just erupted on the national scene, took the lead. It made support for self-consumption part of its electoral program and then, following the election, negotiated a commitment with the People's Party to eliminate barriers to self-consumption. Just days later, in September, even as the process of government formation continued, Citizens proposed a new law (*proposición de ley*) that contained all the provisions proponents of self-consumption had been calling for. The caretaker Rajoy government responded by claiming that the proposed law was inadmissible under the constitution because its adoption would result in a loss of revenue, thereby disrupting the budget for the current year (Ciudadanos 2016a, 2016b; Hernando Fraile and Girauta Vidal 2016; MINETUR 2016).

Undeterred, Citizens vowed to continue to fight for self-consumption following the investiture of the new Rajoy government in October, and, in January 2017, it joined with the other parties to put forward a second proposed law to that end. The new proposal was almost identical to the previous one, but it contained a measure designed to prevent a second government veto: any feature of the legislation that might reduce government revenue would not go into effect until the adoption of the next budget. Otherwise, the government would have three months to take the required action once the law was passed (N.a. 2017; Barrero 2017b; Roca 2017).

Once again, the Rajoy government sought to block consideration of the law by arguing that its adoption would reduce the government's revenue. Nevertheless, because the government no longer enjoyed a majority in the Congress, its veto could be lifted by a majority vote in the Bureau or Presiding Committee (*Mesa*), where the opposition held six seats to the People's Party's three. To everyone's surprise, however, the two Citizens members voted with the government to sustain the veto (MINETUR 2017; Fernández 2017).

What was behind the reversal on the part of Citizens, whereby it effectively vetoed its own proposal? Citizens argued that if it voted against the government, the People's Party would simply take the matter to the Constitutional Court, resulting in a delay of potentially years while the matter was adjudicated. As a result, it would be far more effective to engage in direct negotiations with the government based on the agreement the two parties had reached the previous August. Citizens noted, perhaps somewhat petulantly, that neither of the other major opposition parties, the Socialists and Podemos, had helped it overcome the government's initial veto in October. As the following months went by, however, Citizens' new approach yielded no obvious fruit (Ciudadanos 2017).

*Other Actors*

The paralysis at the national political level did not mean that nothing could be done to promote self-consumption. Spain's decentralized political system reserved considerable powers to the regional governments (*comunidades autónomas* or CCAA), and many of them, including some ruled by the People's Party, seized the opportunity to do so. Although it is not possible to provide an exhaustive list of the many measures taken—they were as diverse as the regions themselves—many of them took the form of direct subsidies, tax credits and deductions, and loans, typically for investment in power generation, especially PV. And in at least one case, Murcia, the regional government sought to pass a law to regulate self-consumption, although it too was blocked at the national level by the People's Party. Not to be outdone, a number of cities and towns, including Madrid and Barcelona, offered financial support, taking advantage of their ability to offer rebates on local real estate and construction taxes (UNEF 2018, 6–7; El Periódico de la Energía 2018a; Barrero 2016).

The period also saw increased efforts and coordination by industry associations and environmental groups supportive of self-consumption. UNEF remained perhaps the leading voice, given the financial stake and interests of its members. But the relatively new, non-partisan Foundation for Renewable Energy emerged as an important protagonist, generating a steady stream of studies and proposals. And in 2017, these organizations joined with nearly 40 other groups to form a united front, the Alliance for Self-Consumption, to promote their common interest (Alianza por el Autoconsumo 2017).

*Legal Efforts*

A variety of actors also initiated legal efforts intended to roll back the restrictions on self-consumption imposed by the Rajoy government, although with mixed results. In May 2017, the Constitutional Court (*Tribunal Constitucional*) issued a ruling that modified or annulled several articles of *Real Decreto* 900/2015 in a case brought by the government of Catalonia. Most importantly, the court struck down the prohibition on shared self-consumption, and it put the regions in charge of registering installations (Tribunal Constitucional 2017).

An October 2017 ruling by the Supreme Court (*Tribunal Supremo*), however, favored the government. In this case, the National Association of Renewable Energy Producers and Investors (*Asociación Nacional de Productores e Inversores de Energías Renovables* or Anpier) claimed that two of the charges imposed by *Real Decreto* 900/2015 violated the 2013 Electricity Sector Law by making self-consumers pay more than traditional power customers. But the court rejected the appeal and also found that the exemption of co-generation from the charges amounted to discriminatory behavior (Tribunal Supremo 2017).

Finally, in April 2018, the Constitutional Court ruled that the People's Party and Citizens could not block consideration by Congress of the law proposed the previous year if it did not affect the current government budget. This ruling prompted the opposition parties, less Citizens this time, to revive their proposal, updated to reflect the previous court rulings, the following month. Before the proposal could be processed, however, an even bigger political development would set the stage for breaking the logjam at the national level (Ojea 2018a; Fernández 2018).

# Reversal of Fortune Under the Sánchez Government

The prospects for self-consumption suddenly brightened in mid-2018, when the Socialists took the reins of power following the successful vote of no confidence against the Rajoy government. In the fall, the new Sánchez government proposed and the Congress approved a set of urgent measures, *Real Decreto-ley* 15/2018, that removed the legal restrictions and obstacles imposed by the 2013 Electricity Sector Law. Then, in May 2019, the government adopted a more detailed set of regulations, *Real Decreto* 244/2019, which formally replace the controversial *Real Decreto*

900/2015. Although a number of technical issues remained to be worked out, the stage was finally set for the full realization of Spain's potential for self-consumption.

### *Initial Steps*

Ironically, the Rajoy government took a first, albeit very modest, step toward promoting self-consumption on the very day the Congress began debate on the no confidence motion. On May 31, it introduced a draft regulation on access and connection to the grid that made changes in the self-consumption regulation to reflect the previous year's ruling of the Constitutional Tribunal. Nevertheless, the measure offered no relief from the special charges and only recognized a limited right to shared self-consumption (N.a. 2018).

Then, in mid-June, following the change in government, the Congress took up the issue of the refusal of the Congress's Bureau to allow consideration of the long proposed draft law. This time, Citizens voted with the other parliamentary groups to allow the proposal to move forward, over the opposition of the People's Party. As the legislative session wound down, in anticipation of the traditional summer recess, however, it became clear that nothing would happen before September (Ojea 2018b).

In the meantime, self-consumption received a boost from the European Union, which was completing the new directive on renewable energy. In mid-June, negotiators for the Commission, Council, and Parliament reached agreement on a final draft, which would be formally approved in December. The draft directive called for empowering renewables self-consumers to generate, consume, store, and sell electricity and admonished that "Renewables self-consumers should not face discriminatory or disproportionate burdens or costs and should not be subject to unjustified charges" (Directive (EU) 2018/2001).

### Real Decreto-ley *15/2018*

When the Congress reconvened, however, work on new legislation to promote self-consumption was overshadowed by a more pressing concern: stubbornly high electricity prices. As a result, the government focused its efforts on developing a set of urgent measures to protect consumers while accelerating the transition to a carbon-free economy.

The shift in focus nevertheless also represented an opportunity for proponents of self-consumption, since approval of such an emergency package was much easier to obtain. And there was no time to lose. In September, the Supreme Court had once again ruled in favor of the special charges on self-consumed electricity, this time in a case brought by UNEF, underscoring the need for new legislation if they were ever to be lifted (Sevillano 2018).

Thus, the final version of the package of urgent measures, which was issued by the government in early October as *Real Decreto-ley* 15/2018 and subsequently validated by a majority of the Congress, included an entire section (*Título*) on self-consumption. Although the inclusion of this section was largely justified on the grounds that self-consumption would help to lower electricity costs, both directly and indirectly by reducing demand in the wholesale market, the preamble also noted the advantages traditionally attributed to it, such as increased energy independence and lower greenhouse gas emissions (Real Decreto-ley 15/2018).

Indeed, the section closely followed the proposed law that had been in circulation for two years, supplemented by some additional material. As such, it modified the 2013 Electricity Sector Law and the 2015 self-consumption regulation in all the ways that had been demanded. It exempted self-consumed electricity from any type of charge or tariff. It recognized the right to shared self-consumption. It simplified administrative procedures and technical requirements, especially for small installations. It provided for the development of simplified compensation mechanisms for installations up to 100 kW. And it greatly reduced the potential fines that could be levied for infractions.

## Real Decreto *244/2019*

The adoption of *Real Decreto-ley* 15/2018 was widely praised by the renewable sector. But it was too early to celebrate. A number of administrative and technical details remained to be worked out via implementing regulation. These included the rules for access and connection to the grid for new installations, the precise mechanism of compensation for excess production fed into the grid, the organization of the administrative register, and the terms for sharing self-consumption. As a result, according to press reports, the sector remained paralyzed (Ojea 2018c; El Periódico de la Energía 2018b).

Initially, it was expected that at least some of these issues would be addressed in the long-awaited comprehensive regulation on access and connection to the grid. But as the prospect for the rapid approval of such a measure became increasingly doubtful, the government decided to focus on developing specific regulation for self-consumption. By then, however, some two months had gone by and it was necessary to proceed with haste if the three-month deadline set in *Real Decreto-ley* 15/2018 were to be met (Roca 2018).

The government issued a first draft of the needed regulations at the end of January. Because of the number and complexity of the issues to be addressed, the document covered more than 60 pages and was accompanied by an impact analysis of some 50 pages. Given the urgency of the situation, however, the required public comment period was limited to less than two weeks. Nevertheless, a number of stakeholders, including the Alliance for Self-Consumption, were able to provide feedback before the deadline passed, and a total of 25, including regional governments, associations, utilities, and others, submitted comments to the CNMC as it prepared its own required report (MITECO 2019a, 2019b; Alianza por el Autoconsumo 2019a).

Overall, the CNMC's evaluation, published in late February 2019, was positive. It found that the proposal simplified the modes of self-consumption, administrative processes, and registration, adequately regulated shared (now called collective) self-consumption, confirmed the elimination of any special charges, created a simplified compensation mechanism for excess production, and facilitated the installation of storage devices. But the report also recommended a number of improvements, and it criticized the government for resorting to the expedited review process, which limited the opportunity for constructive feedback (CNMC 2019a).

The final regulation was approved by the government's Council of Ministers in early April, little more than two months after the original version had been circulated. *Real Decreto* 244/2019 was almost identical to the original version in terms of structure, but it contained a number of clarifications and additional details that responded to the feedback that had been received. Of particular importance were the definitions provided for nearby generation facilities (*instalación de producción próxima*) and collective self-consumption (*autoconsumo colectivo*), which determined the scope of the self-consuming unit on the production and consumption sides, respectively (Real Decreto 244/2019).

Perhaps the most significant development since self-consumption had become a prominent issue concerned the form compensation for excess production fed into the grid would take. The original draft regulation prepared in late 2011 as well as much subsequent discussion focused on the mechanism of net metering, under which the volume of electricity exported would be compensated for up to an equal amount imported without regard to the price at any given moment. As long as the more fundamental obstacle of special charges on self-consumed electricity stood in the way, the details of compensation had received little public attention.

As work on the final regulation began in late 2018 following the adoption of *Real Decreto-ley* 15/2018, however, the government's renewable energy institute (*Instituto para la Diversificación y Ahorro de la Energía* or IDAE) began to express misgivings. In particular, it argued that net metering was not economically efficient and promoted overbuilding of capacity. Instead, it advocated for the approach known as net billing (*facturación neta*). Excess electricity would be sold to the grid at the wholesale market rate prevailing at the time, and any purchases from the grid would be subject to the same price, including access fees, as paid by ordinary consumers. As a result, self-consumers would be incentivized to maximize their use of the electricity they produced (Ojea 2018d).

Notwithstanding the progress represented by the approval of *Real Decreto* 244/2019, a number of further details remained to be worked out. Some concerned how to reduce the administrative barriers to connecting self-consumers to the grid while discouraging speculation in permits, which had been a growing problem. Others concerned the modalities of collective self-consumption. Under the regulation, shares of electricity were determined by static coefficients, which prevented one neighbor from using more than another and could lead to unnecessary exports to the grid. Thus, proponents of collective self-consumption called for making it more flexible using dynamic measures as well as extending it to the medium tension grid and loosening the criteria for what counted as a nearby installation. Not least important were those missing details that concerned the compensation mechanism, which would be a complicated process involving distributors, marketers, and regional governments as well as the self-consumers themselves (Alianza por el Autoconsumo 2019b; Colóm 2019).

Further progress was hampered, however, by the fact that Spain almost immediately held national elections at the end of April and the Sánchez government was reduced to a caretaker role while negotiations—and eventually a second election—to form a new government took place. Thus, it was not until the end of the year that the details of the compensation process were worked out, when the CNMC published two resolutions and the acting Secretary of State for Energy issued a resolution of nearly 300 pages. Self-consumers could finally start receiving compensation in March 2020, or nearly a year after *Real Decreto* 244/2019 was adopted (CNMC 2019b, 2019c; MITECO 2019c).

*Impact*

What impact would these persistent and ultimately successful efforts to promote self-consumption actually have? One should first note that the new laws and regulations alone would not determine the amount of investment. Especially without any significant subsidies, the fate of self-consumption would depend on the cost of building new installations, and these continued to fall. According to a February 2019 UNEF estimate, the cost of self-consumption facilities had declined by 80% over the previous 10 years (UNEF 2019a). In the absence of any special charges on self-consumed electricity, it now seemed possible to amortize an industrial installation in as little as four to six years (Roca 2020).

As a result, investment in self-consumption was already beginning to take off, notwithstanding the obstacles that had been placed by *Real Decreto* 900/2015. Starting at a modest 22 MW in 2014, deployment had roughly doubled each year, reaching 236 MW in 2018 (see Fig. 6.1). By the time that *Real Decreto* 244/2019 was adopted in April 2019, a number of companies were already offering services to those interested in self-consumption, including financing, permitting, construction, operation, and maintenance (Granados 2019).

Based on these trend lines, UNEF variously estimated that the amount of capacity intended for self-consumption would grow by 300–600 MW per year and would constitute roughly 10–20% of all new PV capacity. Similarly, the consulting firm Deloitte calculated that there was a potential for between 5000 and 6500 MW in self-consumption over the next decade. And in early 2020, the new Secretary of State for Energy prognosticated that deployment would double again, reaching 1 GW that year (UNEF 2019b, 9 and 79; Energías Renovables 2019; Deloitte 2018; El Periódico de la Energía 2020).

In contrast to the argument that self-consumption would empower small participants in the electrical system, however, the Deloitte analysis

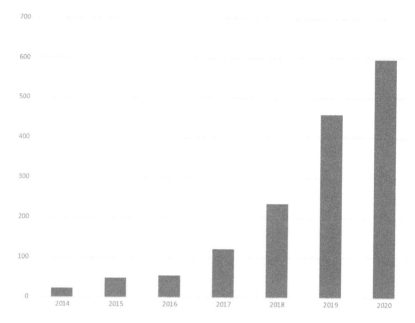

**Fig. 6.1** New self-consumption generating capacity (MW) (*Sources* UNEF 2020, 79, and Roca 2021)

found that the greatest potential (40%) resided in the industrial sector followed by nearly 20% in the service sector and just under 25% divided between small and medium businesses and multi-story residential buildings, figures that were largely confirmed in 2019. The time required to amortize a small business installation (4–6 years) would also be much shorter than that for a residential application (11–13 years). To facilitate the introduction of so much distributed generation, moreover, as much as 14–15 billion euros of investment in modernization and digitalization of the grid might be necessary (Deloitte 2018; Ojea 2020; Monforte 2019).

## Conclusion

Perhaps the biggest puzzle raised by this episode in the politics of renewable power in Spain is the degree of conflict that existed. Nowhere else have we seen as much disagreement between the major political parties over a matter of policy. At least partly as a result, the takeoff of self-consumption was delayed for the better part of a decade, from as early as 2012 until 2019.

Some of the differences can be attributed to party ideology. The People's Party, on the one hand, and the other major parties, on the other, simply assigned different priorities to the potentially competing objectives of reducing the tariff deficit and promoting self-consumption, respectively. These differences might in turn be attributed to different degrees of emphasis on more general fiscal and environmental-social issues, which came into greater conflict as a result of the financial and economic crisis.

These ideological predispositions were reinforced by the politics of interest groups, which enjoyed varying degrees of access to and influence with the different parties. The major utilities were threatened by the possibility that growing numbers of consumers would generate their own electricity, and they seemed to have the ear of the Rajoy government (Mir-Artigues 2013; Barrero 2013b). Meanwhile, the Socialists and the emerging Podemos (We can) and Citizens parties were more inclined to listen to the renewable power industry and its supporters, which in any case became increasingly vocal and well organized. Both UNEF and the Renewable Energy Foundation maintained a steady drumbeat of advocacy for self-consumption, culminating in their organization of most if not all interested parties in the Alliance for Self-Consumption.

Although the People's Party lost its Congressional majority at the end of 2015 and was forced to rule as a minority government thereafter, it was able to delay the development of favorable legislation and regulation for several more years thanks to a combination of procedural mechanisms. One was the requirement that new legislation not reduce the revenues in the current budget agreement. And when that was challenged, the party was able to use its majority, when supported by Citizens, in the Congress's Bureau to block parliamentary consideration, at least until that option was removed by the Constitutional Court.

Finally, some have chalked up the People's Party's resistance largely to personalities. In particular, policy toward self-consumption was substantially shaped by first Alberto Nadal, who served as Secretary of State for Energy from December 2011 until November 2016, and then his brother, Álvaro Nadal, who served as Minister of Energy, Tourism, and Digital Agenda from November 2016 until the end of the Rajoy government, and these were two of the most outspoken proponents of the backup tax (Albasolar 2016; Monforte and de Miguel 2017).

## Notes

1. *Ley* 54/1997 defined a self-producer as one that consumes at least 30 percent of the electricity generated by a facility of up to 25 MW and at least 50 percent for a facility of more than 25 MW (Art. 9, 1b).
2. Interestingly, the People's Party, which lost its majority in the Congress in the general elections of December 2015 and June 2016 and had to form a minority government, never adopted the additional regulation needed to calculate the variable charge on self-consumed electricity. As a result, the charge was never collected. Deployment suffered any way, however, as potential investors assumed they would eventually have to pay it (Rodríguez Labastida 2018).

## References

Acuerdo. 2015. "Acuerdo de medidas urgentes de fomento del autoconsumo eléctrico" (24 Feb.). https://drive.google.com/file/d/0B3HtyHIhg0gnYVFFMGVVbGxGQ1E/view?pref=2&pli=1.

Albasolar. 2016. "Alberto Nadal insulta a los autoconsumidores llamándoles 'depredadores'" (18 March). https://albasolar.es/alberto-nadal-insulta-a-los-autoconsumidores-llamandoles-depredadores/.

Alianza por el Autoconsumo. 2017. "Alianza por el autoconsumo Manifiesto: Con el autoconsumo ganamos todos" (11 May). https://fundacionrenovables.org/wp-content/uploads/2017/05/Alianza-por-el-Autoconsumo.-Manifiesto-Final-11-05-2017.pdf.

———. 2019a. "Propuesta de real decreto por el que se regulan las condiciones administrativas, técnicas y económicas del autoconsumo" (8 Feb.). https://docs.google.com/document/d/1NEhbZzTkVHu_vi-qNq8Hf-D5ovBt3pIt3u07zYIPqss/edit.

———. 2019b. "Alianza por el Autoconsumo espera que el desarrollo del nuevo Real Decreto haga que el autoconsumo sea masivo en España" (10 April). https://fundacionrenovables.org/notas/alianza-por-el-autoconsumo-espera-que-el-desarrollo-del-nuevo-real-decreto-haga-que-el-autoconsumo-sea-masivo-en-espana/.

ASIF (Asociación de la Industria Fotovoltaica). 2009. "ASIF propone modificar la regulación para que los consumidores puedan producir y consumir su propia electricidad fotovoltaica" (23 Nov.). https://www.construible.es/2009/11/23/asif-propone-modificar-la-regulacion-para-que-los.

Barrero, Antonio. 2013a. "El autoconsumo beneficia también a los que no son autoconsumidores." *Renewable Energy Magazine* (17 Sept.). https://www.renewableenergymagazine.com/panorama/el-autoconsumo-beneficia-tambien-a-los-que-20130917.
———. 2013b. "¿Quién le ha escrito a Nadal el real decreto de autoconsumo?" *Energías Renovables* (25 July). https://www.energias-renovables.com/fotovoltaica/quien-le-ha-escrito-a-nadal-20130724.
———. 2016. "Ayuntamientos y Comunidades Autónomas se enfrentan al nuevo ministro de Energía." *Renewable Energy Magazine* (2 Dec.). https://www.renewableenergymagazine.com/panorama/esta-es-la-proposicion-de-ley-para-20170125/panorama/ayuntamientos-y-comunidades-autonomas-se-enfrentan-al-20161202.
———. 2017a. "Cinco años sin impuesto al sol." *Energías Renovables* (4 Jan.). https://www.energias-renovables.com/fotovoltaica/cinco-anos-sin-impuesto-al-sol-20170104.
———. 2017b. "El autoconsumo, a 90 días vista." *Energías Renovables* (25 Jan.). https://www.energias-renovables.com/fotovoltaica/el-autoconsumo-solar-fotovoltaico-20170125.
BOCG (Bulletin Oficial de las Cortes Generales). 1991. *Plan Energético Nacional 1991–2000* (13 Sept.). http://www.congreso.es/public_oficiales/L4/CONG/BOCG/E/E_169.PDF.
Castillo, Manuel, Marta Victoria, and Iñigo Ramiro. 2015. "Crónica de un nefasto proyecto para regular el autoconsumo" (30 Aug.). http://observatoriocriticodelaenergia.org/?p=2072.
Ciudadanos. 2016a. "350 Soluciones Para Cambiar España a Mejor" (n.d.). https://d3cra5ec8gdi8w.cloudfront.net/uploads/documentos/2016/06/13/documentos_ciudadanos_b358e296.pdf.
———. 2016b. "Proposición de Ley de medidas para el fomento del autoconsumo eléctrico" (9 Sept.). http://www.congreso.es/public_oficiales/L12/CONG/BOCG/B/BOCG-12-B-27-1.PDF.
———. 2017. "Notas de Prensa" (22 March). https://www.ciudadanos-cs.org/prensa/rodriguez-hemos-abierto-una-mesa-de-negociacion-con-el-gobierno-para-desbloquear-el-autoconsumo-como-recoge-el-acuerdo-de-investidura/9562.
CNC (Comisión Nacional de la Competencia). 2013. "IPN 103/13 Anteproyecto de Ley del Sector Eléctrico" (9 Sept.). https://www.cnmc.es/sites/default/files/424533_7.pdf.
CNE (Comisión Nacionao de Energía). 2012. *Informe 3/2012 de la CNE Sobre la Propuesta de Real Decreto por el que Se Establece la Regulación de las Condiciones Administrativas Técnicas y Económicas de la Modalidad de Suministro de Energía Eléctrica Con Balance Neto* (28 March). https://www.cnmc.es/en/node/357758.

———. 2013a. *Informe 19/2013 de la CNE Sobre la Propuesta de Real Decreto or el que Se Establece la Regulación de las Condiciones Administrativas, Técnicas y Económicas de las Modalidades de Suministro de Energía Eléctrica con Autoconsumo y de Producción con Autoconsumo* (4 Sept.). https://www.cnmc.es/sites/default/files/1552840_7.pdf.

———. 2013b. *Informe 16/2013 de la CNE Sobre el Anteproyecto de Ley del Sector Eléctrico* (31 July). https://www.cnmc.es/sites/default/files/1552718_8.pdf.

CNMC (Comisión Nacional de los Mercados y la Competencia). 2015. *Informe Sobre el Proyecto de Real Decreto por el que Se Establece la Regulación de las Condiciones Administrativas, Técnicas y Económicas de las Modalidades de Suministro de Energía Eléctrica Con Autoconsumo y de Producción Con Autoconsumo.* IPN/DE/011/15 (8 July). https://www.cnmc.es/sites/default/files/1491290_8.pdf.

———. 2019a. *Acuerdo por el que se Emite Informe sobre laPpropuesta de Real Decreto por el que se Regulan las Condiciones Administrativas Técnicas y Económicas del Autoconsumo.* IPN/CNMC/005/19 (21 Feb). https://www.cnmc.es/en/node/373814.

———. 2019b. "Resolución de 7 de noviembre de 2019, de la Comisión Nacional de los Mercados y la Competencia, por la que se aprueba la adaptación del procedimiento de operación 14.8 'Sujeto de liquidación de las instalaciones de producción y de las instalaciones de autoconsumo' y del procedimiento de operación 14.4 'Derechos de cobro y obligaciones de pago por los servicios de ajuste del sistema' al Real Decreto 244/2019, de 5 de abril" (7 Nov.). https://www.boe.es/eli/es/res/2019/11/07/(2).

———. 2019c. *Resolución por la que Se Aprueba el Formato de los Ficheros de Intercambio de Información Entre Comunidades y Ciudades Con Estatuto de Autonomía y Distribuidores Para la Remisión de Información Sobre el Autoconsumo de Energía Eléctrica.* INF/DE/092/19 (13 Nov.). https://www.cnmc.es/sites/default/files/2750205_11.pdf.

Colóm, María. 2019. "Evolución del Autoconsumo en España y Sus Perspectivas de Futuro" (21 Oct.). http://apiem.org/manual-del-instalador/item/ponencia-de-maria-colom-directora-de-politicas-energeticas-de-unef-evolucion-del-autoconsumo-en-espana-y-sus-perspectivas-de-futuro.

Deloitte. 2018. "La contribución de las redes eléctricas a la descarbonización de la generación eléctrica y la movilidad" (3 Dec.). https://www2.deloitte.com/es/es/pages/strategy/articles/contribucion-redes-electricas-descarbonizacion.html.

Directive 2009/28/EC of the European Parliament and of the Council of 23 April 2009 on the promotion of the use of energy from renewable sources and amending and subsequently repealing Directives 2001/77/EC and 2003/30/EC. *Official Journal of the European Union*

(5 June 2009). https://eur-lex.europa.eu/LexUriServ/LexUriServ.do?uri=OJ:L:2009:140:0016:0062:EN:PDF.

Directive (EU) 2018/2001 of the European Parliament and of the Council of 11 December 2018 on the promotion of the use of energy from renewable sources. *Official Journal of the European Union* (21 Dec.). https://eur-lex.europa.eu/legal-content/EN/TXT/PDF/?uri=CELEX:32018L2001&from=EN.

El País. 2015. "Industria creará un gravamen para el autoconsumo de luz con baterías" (8 June). https://elpais.com/economia/2015/06/08/actualidad/1433781909_066749.html.

El Periódico de la Energía. 2015. "El Gobierno podría aplazar el decreto de autoconsumo para septiembre para 'mejorarlo'" (22 July). https://elperiodicodelaenergia.com/el-gobierno-podria-aplazar-el-decreto-de-autoconsumo-para-septiembre-para-mejorarlo/.

———. 2018a. "PP, PSOE, Podemos y C's aceptan tramitar en el Congreso la ley de autoconsumo de la Asamblea de Murcia" (9 Oct.). https://elperiodicodelaenergia.com/pp-psoe-podemos-y-cs-aceptan-tramitar-en-el-congreso-la-ley-de-autoconsumo-de-la-asamblea-de-murcia/.

———. 2018b. "APPA pide que se agilice la aprobación del reglamento para el autoconsumo y que haya incentivos fiscales" (29 Nov.). https://elperiodicodelaenergia.com/appa-pide-que-se-agilice-la-aprobacion-del-reglamento-para-el-autoconsumo-y-que-haya-incentivos-fiscales/.

———. 2020. "El Gobierno sacará las nuevas subastas de renovables «lo antes posible» y dará nuevas ayudas a través del IDEA" (5 Feb.). https://elperiodicodelaenergia.com/el-gobierno-sacara-las-nuevas-subastas-de-renovables-lo-antes-posible-y-dara-nuevas-ayudas-a-traves-del-idae/.

EnerAgen. n.d. "¿Qué es el autoconsumo?" http://www.autoconsumoaldetalle.es/que-es-al-autoconsumo/.

Energías Renovables. 2015a. "El RD de autoconsumo enviado a la CNMC mantiene el impuesto al sol" (5 June). https://www.energias-renovables.com/fotovoltaica/el-rd-de-autoconsumo-enviado-hoy-a-20150605.

———. 2015b. "Los partidos políticos se comprometen a eliminar el 'impuesto al sol'" (4 Nov.). https://www.energias-renovables.com/fotovoltaica/los-partidos-politicos-se-comprometen-a-eliminar-20151104.

———. 2019. "El autoconsumo instalará en España entre 450 y 600 megavatios de potencia cada año" (4 March). https://www.energias-renovables.com/fotovoltaica/el-autoconsumo-instalara-en-espana-entre-450-20190302.

Europa Press. 2015a. "Industria cambia el 'impuesto al sol' por una nueva tasa al autoconsumo" (5 June). https://www.eldiario.es/economia/industria-autoconsumidor-producido-nuevo-cargo_1_2638259.html.

Fernández, Marta. 2017. "Ciudadanos respalda el veto del PP a la ley contra el 'impuesto al sol.'" *el Boletin* (14 March). https://www.elboletin.com/economia/146868/ciudadanos-pp-veto-contra-impuesto-sol.html.

———. 2018. "La oposición (sin Ciudadanos) lleva de nuevo al Congreso una iniciativa para tumbar el 'impuesto al sol.'" *el Boletin* (10 May). https://www.elboletin.com/noticia/162920/economia/la-oposicion-sin-ciudadanos-lleva-de-nuevo-al-congreso-una-iniciativa-para-tumbar-el-impuesto-al-sol.html.

Granados, Óscar. 2019. "El sol brilla más desde este año." *El País* (21 May). https://elpais.com/elpais/2019/05/20/actualidad/1558365184_918645.html.

Greenpeace. 2015. "Greenpeace rechaza el impuesto al sol con el que el Ministerio de Industria quiere penalizar el autoconsumo" (8 June). http://archivo-es.greenpeace.org/espana/es/news/2015/Junio/Greenpeace-rechaza-el-impuesto-al-sol-con-el-que-el-Ministerio-de-Industria-quiere-penalizar-el-autoconsumo/.

Hernando Fraile, Rafael, and Juan Carlos Girauta Vidal. 2016. "150 Compromisos para Mejorar España" (28 Aug.). http://canarias.ciudadanos-cs.org/wp-content/uploads/sites/76/2016/10/pacto-150medidas.pdf.

IDAE (Instituto para la Diversificación y Ahorro de la Energía). 1999. *Plan de Fomento de las Energías Renovables en España* (Dec.). https://www.idae.es/uploads/documentos/documentos_4044_PFER2000-10_1999_1cd4b316.pdf.

———. 2005. *Plan de Energías Renovables en España (PER) 2005–2010* (Aug.). https://www.idae.es/publicaciones/plan-de-energias-renovables-en-espana-2005-2010.

———. 2010. *Plan de Acción Nacional de Energías Renovables de España (PANER) 2011–2020* (30 June). https://www.mincotur.gob.es/energia/desarrollo/EnergiaRenovable/Documents/20100630_PANER_Espanaversion_final.pdf.

———. 2011. *Plan de Energías Renovables (PER) 2011–20* (11 Nov.). http://www.idae.es/file/9712/download?token=6MoeBdCb.

Jacobs, David. 2012. *Renewable Energy Policy Convergence in the EU: The Evolution of Feed-in Tariffs in Germany, Spain, and France*. London and New York: Routledge.

Kenning, Tom. 2015a. "Spanish Fines for Self-Consumption Double That for Leaking Radioactive Waste." *PV Tech* (16 June). https://www.pv-tech.org/news/spain_proposes_sun_tax_on_storage_of_self_consumption_solar.

———. 2015b. "'Record' Petition Against Spanish Tax on Solar Self-Consumption and Storage." *PV Tech* (1 July). https://www.pv-tech.org/news/record_petition_against_spanish_tax_on_solar_self_consumption_and_storage.

———. 2015c. "Spanish Government Under Fire After Approving 'Sun Tax.'" *PV Tech* (12 Oct.). https://www.pv-tech.org/news/spains_sun_tax_approved_to_avoid_development_of_pv.

———. 2016. "Spanish Parliament Rallies Against Punitive 'Sun Tax.'" *PV Tech* (9 March). https://www.pv-tech.org/news/spanish-sun-tax-on-brink-of-removal.

Ley 82/1980, de 30 de diciembre, sobre conservación de energía. 1980. https://www.boe.es/eli/es/l/1980/12/30/82.

Ley 54/1997, de 27 de noviembre, del Sector Eléctrico. 1997. https://www.boe.es/buscar/doc.php?id=BOE-A-1997-25340.

Ley 24/2013, de 26 de diciembre, del Sector Eléctrico (26 Dec. 2013). https://www.boe.es/eli/es/l/2013/12/26/24.

MINCOTUR (Ministerio de Industria, Comercio y Turismo). 2015. "The Government adopted the regulation of electricity consumption" (8 Oct.). https://www.mincotur.gob.es/en-US/GabinetePrensa/NotasPrensa/2015/Paginas/20151009-rd-autoconsumo.aspx.

MINETUR (Ministerio de Industria, Energía y Turismo). 2013. "Propuesta de Real Decreto" (18 July).

———. 2016. "Informe de la secretaria de estado de energia" (13 Oct.). https://elperiodicodelaenergia.com/wp-content/uploads/2017/01/Informe-Gobierno-Autoconsumo-propuesta-Ciudadanos.pdf.

———. 2017. "Informe de la secretaria de estado de energia" (9 March).

MITECO (Ministerio para la Transición Ecológica). 2019a. "Propuesta de real decreto por el que se regulan las condiciones administrativas, técnicas y económicas del autoconsumo" (Jan.). https://www.miteco.gob.es/es/ministerio/servicios/participacion-publica/2019_01_29_rd_autoconsumo_final_v2_tcm30-486582.pdf.

———. 2019b. "Memoria del análisis de impacto normativo de la propuesta de real decreto por el que se regulan las condiciones administrativas, técnicas y económicas del autoconsumo" (29 Jan.). https://www.miteco.gob.es/es/ministerio/servicios/participacion-publica/2019_01_29_main_rd_autoconsumo_final_tcm30-486581.pdf.

———. 2019c. "Resolución de 11 de diciembre de 2019, de la Secretaría de Estado de Energía, por la que se aprueban determinados procedimientos de operación para su adaptación al Real Decreto 244/2019, de 5 de abril, por el que se regulan las condiciones administrativas, técnicas y económicas del autoconsumo de energía eléctrica" (20 Dec.). https://www.boe.es/eli/es/res/2019/12/11/(3).

Mir-Artigues, Pere. 2012. *Economía de la generación eléctrica solar: La regulación fotovoltaica y solar termoeléctrica en España*. Civitas/Thomson Reuters.

———. 2013. "The Spanish Regulation of the Photovoltaic Demand-side Generation." *Energy Policy* 63 (Dec.): 664–673.

Mir-Artigues, Pere, Pablo del Río, and Emilio Cerdá. 2018. "The Impact of Regulation on Demand-Side Generation." *Energy Policy* 121 (Oct.): 286–291.

Monforte, Carmen. 2019. "Las eléctricas estiman 15.000 millones de inversión para impulsar el autoconsumo." *CincoDías* (17 April). https://cincodias.elpais.com/cincodias/2019/04/16/companias/1555437557_454461.html.

Monforte, Carmen, and Bernardo de Miguel. 2017. "Nadal arremete ante la UE contra el autoconsumo y lo califica de 'oneroso.'" *CincoDías* (29 April). https://cincodias.elpais.com/cincodias/2017/04/28/companias/1493406355_589447.html.

N.a. 2017. "Proposición de ley de medidas para el fomento del autoconsumo eléctrico" (25 Jan.). https://www.energias-renovables.com/ficheroenergias/Proposicion_de_Ley_para_el_Fomento_del_Autoconsumo_Electrico.pdf.

———. 2018. "Propuesta de real decreto de acceso y conexión a las redes de transporte y distribución" (n.d.). https://energia.gob.es/es-ES/Participacion/Documents/rd-acceso-conexion-redes-transporte/RD-Conexion-Acceso.pdf.

Ojea, Laura. 2018a. "Ciudadanos da una segunda oportunidad al PP para mantener el veto a la eliminación del 'impuesto al sol.'" *El Periódico de la Energía* (30 April). https://elperiodicodelaenergia.com/ciudadanos-da-una-segunda-oportunidad-al-pp-para-mantener-el-veto-a-la-eliminacion-del-impuesto-al-sol/.

———. 2018b. "La ley que acabará con el 'impuesto al sol' tendrá que esperar a septiembre." *El Periódico de la Energía* (25 June). https://elperiodicodelaenergia.com/la-ley-que-acabara-con-el-impuesto-al-sol-tendra-que-esperar-a-septiembre/.

———. 2018c. "El extraño parón del autoconsumo en España tras la eliminación del 'impuesto al sol.'" *El Periódico de la Energía* (14 Nov.). https://elperiodicodelaenergia.com/el-extrano-paron-del-autoconsumo-en-espana-tras-la-eliminacion-del-impuesto-al-sol/.

———. 2018d. "Joan Herrera (IDAE): «No se va a poder hacer balance neto en autoconsumo sino factura neta.»" *El Periódico de la Energía* (19 Nov.). https://elperiodicodelaenergia.com/joan-herrera-idae-no-se-va-a-poder-hacer-balance-neto-en-autoconsumo-sino-factura-neta/.

———. 2020. "El autoconsumo remonta en España pese al parón por la pandemia: las expectativas de crecimiento podrían superar a 2019." *El Periódico de la Energía* (21 May). https://elperiodicodelaenergia.com/el-autoconsumo-remonta-en-espana-pese-al-paron-por-la-pandemia-las-expectativas-de-crecimiento-podrian-superar-a-2019/.

Presidencia. 2015. "Referencia del Consejo de Ministros" (9 Oct.). https://www.lamoncloa.gob.es/consejodeministros/referencias/Paginas/2015/refc20151009.aspx.

Real Decreto 2366/1994, de 9 de diciembre, sobre producción de energía eléctrica por instalaciones hidráulicas, de cogeneración y otras abastecidas por recursos o fuentes de energía renovables (31 Dec. 1994). https://www.boe.es/eli/es/rd/1994/12/09/2366.

Real Decreto 1699/2011, de 18 de noviembre, por el que se regula la conexión a red de instalaciones de producción de energía eléctrica de pequeña potencia (18 Nov. 2011). https://www.boe.es/eli/es/rd/2011/11/18/1699/con.

Real Decreto 900/2015, de 9 de octubre, por el que se regulan las condiciones administrativas, técnicas y económicas de las modalidades de suministro de energía eléctrica con autoconsumo y de producción con autoconsumo (10 Oct. 2015). https://www.boe.es/eli/es/rd/2015/10/09/900.

Real Decreto 244/2019, de 5 de abril, por el que se regulan las condiciones administrativas, técnicas y económicas del autoconsumo de energía eléctrica (5 April 2019). https://www.boe.es/eli/es/rd/2019/04/05/244.

Real Decreto-ley 13/2012, de 30 de marzo, por el que se transponen directivas en materia de mercados interiores de electricidad y gas y en materia de comunicaciones electrónicas, y por el que se adoptan medidas para la corrección de las desviaciones por desajustes entre los costes e ingresos de los sectores eléctrico y gasista (31 March 2012). https://www.boe.es/eli/es/rdl/2012/03/30/13.

Real Decreto-ley 15/2018, de 5 de octubre, de medidas urgentes para la transición energética y la protección de los consumidores (6 Oct. 2018). https://www.boe.es/eli/es/rdl/2018/10/05/15.

Roca, Marc. 2013. "Spain Hurts Solar with Plan to Penalize Power Producers." *Bloomberg* (1 Aug.). https://www.bloomberg.com/news/articles/2013-08-01/spain-hurts-solar-with-plan-to-penalize-power-producers.

Roca, Ramón. 2015. "El decreto de autoconsumo ya está listo, con todos los cargos incluidos, y a punto de enviarse al Consejo de Estado." *El Periódico de la Energía* (24 July). https://elperiodicodelaenergia.com/el-decreto-de-autoconsumo-ya-esta-listo-con-todos-los-cargos-incluidos-y-a-punto-de-enviarse-al-consejo-de-estado/.

———. 2017. "Todos los partidos (menos el PP) vuelven a unirse este miércoles para tumbar 'el impuesto al sol.'" *El Periódico de la Energía* (24 Jan.). https://elperiodicodelaenergia.com/todos-los-partidos-menos-el-pp-vuelven-a-unirse-este-miercoles-para-tumbar-el-impuesto-al-sol/.

———. 2018. "Ribera no espera al RD de Acceso y Conexión y tramitará otro de urgencia para agilizar el autoconsumo en España." *El Periódico de la Energía* (7 Dec.). https://elperiodicodelaenergia.com/ribera-no-espera-al-rd-de-acceso-y-conexion-y-tramitara-otro-de-urgencia-para-agilizar-el-autoconsumo-en-espana/.

———. 2020. "Los ingenieros industriales reclaman la regulación de la figura del agregador para impulsar la generación distribuida." *El Periódico de la*

*Energía* (4 June). https://elperiodicodelaenergia.com/los-ingenieros-industriales-reclaman-la-regulacion-de-la-figura-del-agregador-para-impulsar-la-generacion-distribuida/.

———. 2021. "El autoconsumo doblega a la pandemia: se instalan en España 596 MW en 2020, un 30% más que un año antes." *El Periódico de la Energía* (28 Jan.). https://elperiodicodelaenergia.com/el-autoconsumo-doblega-a-la-pandemia-se-instalan-en-espana-596-mw-en-2020-un-30-mas-que-un-ano-antes/.

Rodríguez Labastida, Roberto. 2018. "Is Distributed Generation in Spain Waking Up from Its Long Nap?" (8 March). https://guidehouseinsights.com/news-and-views/is-distributed-generation-in-spain-waking-up-from-its-long-nap.

Sáenz de Miera, Gonzalo. 2012. "Análisis del autoconsumo en el marco del sector eléctrico español" (23 May). https://www.yumpu.com/es/document/read/33114686/analisis-del-autoconsumo-en-el-marco-del-sector-caloryfriocom.

Sevillano, Elena. 2018. "El Supremo vuelve a avalar el 'impuesto al sol' del Gobierno Rajoy." *El País* (9 Sept.). https://elpais.com/economia/2018/09/07/actualidad/1536348758_003803.html.

Transparency International España. 2014. *An Evaluation of Lobbying In Spain: Analysis and Proposals* (Dec.).

Tribunal Constitucional. 2017. "Sentencia 68/2017, de 25 de mayo 2017" (1 July). https://www.boe.es/buscar/doc.php?id=BOE-A-2017-7644.

Tribunal Supremo. 2017. "El Supremo avala el Real Decreto autoconsumo eléctrico y rechaza que exista un 'impuesto al sol.'" *Notícias Jurídicas* (19 Oct.). http://noticias.juridicas.com/actualidad/noticias/12426-el-supremo-avala-el-real-decreto-autoconsumo-electrico-y-rechaza-que-exista-un-quot;impuesto-al-solquot.

UNEF (Unión Española Fotovoltaica). 2013a. *Informe Anual 2013*. https://unef.es/wp-content/uploads/dlm_uploads/2013/09/memo-unef_2013.pdf.

———. 2013b. "El Ministerio de Industria impide el autoconsumo de electricidad" (19 July). http://www.energiza.org/index.php?option=com_k2&view=item&id=608:unef-'el-ministerio-de-industria-impide-el-autoconsumo-de-electricidad'.

———. 2014. *Informe Anual 2014*. https://unef.es/wp-content/uploads/downloads/2014/11/MEMO-UNEF_2014.pdf.

———. 2016. *Informe Anual 2016*. https://unef.es/wp-content/uploads/dlm_uploads/2016/08/Informe-Anual-UNEf-2016_El-tiempo-de-la-energia-solar-fotovoltaica.pdf.

———. 2018. *Informe Anual* 2018. https://unef.es/wp-content/uploads/dlm_uploads/2018/09/memo_unef_2018.pdf.

———. 2019a. "La energía fotovoltaica en España avanza de forma decidida en 2018, con un crecimiento del 94% de la potencia instalada" (5 Feb.). https://unef.es/2019/02/la-energia-fotovoltaica-en-españa-avanza-de-forma-decidida-en-2018-con-un-crecimiento-del-94-de-la-potencia-instalada/.
———. 2019b. *Informe Anual 2019*. https://unef.es/wp-content/uploads/dlm_uploads/2019/09/memoria_unef_2019-web.pdf.
———. 2020. *Informe Anual 2020*. https://unef.es/informacion-sectorial/informe-anual-unef/.

CHAPTER 7

# Future Prospects for Renewable Power in Spain

## INTRODUCTION

As we have seen in the previous chapters, support for renewable power in Spain—and the resulting growth in renewable generating capacity—has waxed and waned with changing circumstances over the last several decades. This chapter examines what is likely to happen in the years ahead. At this point, the prospects for a favorable environment are generally positive. The left-wing coalition government formed in early 2020 has established ambitious and detailed targets for the growth of renewable capacity through 2030. Attaining those targets, and the underlying decarbonization goals they are intended to serve, however, is not assured. One can identify three broad challenges: (1) actually building sufficient capacity, (2) connecting so much capacity to the grid, and (3) successfully integrating the resulting production into an electricity market that was based on the assumption that power generation could always be adjusted to meet demand. To address these issues, the coalition government has devised—and continues to refine—a number of policies that, taken together, could transform the power system to a greater extent than any time since the liberalization measures of the 1990s.

© The Author(s), under exclusive license to Springer Nature Switzerland AG 2021
J. S. Duffield, *Making Renewable Electricity Policy in Spain*, Environmental Politics and Theory,
https://doi.org/10.1007/978-3-030-75641-3_7

## Motivations

The biggest driver of future efforts to promote renewable power in Spain is likely to be concern about climate change. Spain has played an active role in global efforts to reduce greenhouse gas emissions, as exemplified by its agreement to host, at the last minute, the 2019 international climate summit. Spain quickly signed on to the 2015 Paris climate agreement, and in early 2020, the new coalition government declared a national climate emergency as a first formal step toward introducing a comprehensive package of climate policies, headed by a new Law on Climate Change and the Energy Transition (*Ley de Cambio Climático y Transición Energética* or LCCTE).

The contours of Spain's renewable energy policy, however, will continue to be largely shaped by its membership in the European Union (EU). In 2014, the European Council agreed on a climate and energy framework that included binding targets for 2030 of at least a 40% reduction in greenhouse gas emissions, compared to 1990, and at least 27% of energy consumption from renewable sources (European Council, n.d.). The former target became the EU's contribution under the Paris Agreement. Then, as the decade drew to a close, the EU adopted a comprehensive package of energy legislation, including an updated renewable energy directive that raised the overall target to 32% of energy consumption in 2030 (Directive (EU) 2018/2001).

Nevertheless, these external forces would continue to be complemented by a series of long-standing domestic considerations. Among these were reducing foreign energy dependence, stimulating national industrial development, increasing employment, and promoting local and regional growth. And hardly had the 2020s begun than were the traditional economic imperatives reinforced by the belief that investment in renewable energy could accelerate Spain's recovery from the economic ravages of the coronavirus pandemic.

## Renewable Energy Targets

In response to these pressures, the successive *Partido Socialista Obrero Español* (PSOE) and PSOE-Podemos governments of 2018 and 2020 developed ambitious renewable power targets for 2030. These targets were contained in Spain's Integrated National Energy and Climate Plan (*Plan Nacional Integrado de Energía y Clima* or PNIEC), a first draft of which was issued in early 2019 and which was finalized in early 2020. The PNIEC set an ambitious overall renewable energy target of 42% of final energy consumption by 2030, more than double Spain's target for 2020, with no less than 74% of electricity coming from renewable sources. To achieve the latter goal, the plan called for an increase in the amount of renewable capacity of some 60 gigawatts (GW), with most of that taking the form of solar photovoltaic (PV) and onshore wind.

### *PNIEC Origins and Development*

The development of the PNIEC dates back to at least 2015, when the European Commission called on member states to prepare such plans and provided detailed guidance on how they should be structured. The deadline for the submission of first drafts was originally set for 2017, but this was delayed till the end of 2018, as the Commission's original call was recast as a central component of the EU's proposed regulation on the governance of the European Energy Union, which was not completed until that year (Regulation (EU) 2018/1999). Over the same period, the EU also developed a new renewable energy directive, which set a binding global target for 2030 of at least 32%, a substantial increase over the 2020 goal of 20% (Directive (EU) 2018/2001). Specific national objectives would be set by the EU member states themselves in their individual plans.

Initially, Spain was slow to comply with the planning requirement. As a first step, in the summer of 2017, the center-right government of Mariano Rajoy established a Commission of Experts to analyze scenarios that would lead to the achievement of likely European energy and climate objectives. The Commission's report, issued in April 2018, employed a reference scenario for 2030 in which nearly 30% of final energy consumption, and 62% of electricity, came from renewable sources. To achieve these levels, Spain would need a total of 31 GW of wind and 47.15 GW of solar PV capacity. And in an extreme scenario, involving 47.5 GW of wind and 77 GW of PV, Spain could provide 70% of its power and 33.3% of all energy from renewables (Comisión 2018, 6, 8–9).

Otherwise, however, the Rajoy government made little progress on the plan before it was forced out of office in June 2018, leaving little time for the new government of Pedro Sánchez to meet the end of year deadline. The planning process was further complicated by the need to address thorny political issues such as whether and when to shut down nuclear and coal-fired power plants.

Thus, it was not till February 2019 that the Sánchez government was able to complete the first draft of the PNIEC (MITECO 2019a). But what the draft lacked in timeliness it more than made up for in ambition. As part of a broader strategy to reduce its greenhouse gas emissions by 21% with respect to 1990 levels, Spain would obtain no less than 42% of its final energy consumption and 74% of its electricity from renewable sources by 2030. More impressively, the share of electricity that was renewable would increase even as total power consumption grew, by more than 10%.

To achieve these goals, the amount of renewable generating capacity would have to increase dramatically. From 2020 to 2030, wind capacity would rise from 28.0 GW to 50.3 GW, and some 20 GW of the existing installations would be repowered. PV capacity would grow even more dramatically, from 8.4 GW to 36.9 GW. And even the amount of concentrated solar power (CSP) would rise notably, from 2.3 GW to 7.3 GW. Overall, renewable capacity would more than double, by some 55 GW.

For the next month and a half, the draft PNIEC was subject to public consultation, and the government received more than 12,000 sets of comments, reflecting broad and intense interest in the subject (Roca 2019a). Overall, the renewable industry expressed considerable satisfaction with the plan. The draft was also forwarded to the European Commission, which provided its assessment and recommendations in June. The Commission welcomed Spain's level of ambition and, with regard to renewables, mainly asked Spain to provide additional details about how it would promote them (EC 2019).

The prolonged period of caretaker government following the May 2019 national election slowed the adoption of a second draft. But once the new PSOE-Podemos government was formed in early 2020, it quickly moved to approve the plan, along with the required economic and social impact and environmental strategy studies, by late January, less than a month after the EU deadline. Although the greenhouse gas reduction goal was increased to 23%, the targets for renewable power remained the same, with one notable exception: that for solar photovoltaic rose by 2.3 GW, to a new total of 39.2 GW (MITECO 2020a) (Table 7.1).

Table 7.1 PNIEC capacity targets (MW)

| | 2020 | 2025 | 2030 |
|---|---|---|---|
| Wind | 28,033 | 40,633 | 50,333 |
| Solar PV | 9071 | 21,713 | 39,181 |
| CSP | 2303 | 4803 | 7303 |
| Hydroelectric | 14,109 | 14,359 | 14,609 |
| Pumped (mixed) | 2687 | 2687 | 2687 |
| Pumped (pure) | 3337 | 4212 | 6837 |
| Biogas | 211 | 241 | 241 |
| Other renewables | 0 | 40 | 80 |
| Biomass | 613 | 815 | 1408 |
| Storage | 0 | 500 | 2500 |
| Combined cycle | 26,612 | 26,612 | 26,612 |
| Nuclear | 7399 | 7399 | 3181 |
| Coal | 7897 | 2165 | 0 |

*Source* MITECO (2020a, 46)

## *Longer-Term Objectives*

Even before implementation of the PNIEC began in earnest, Spain began to set even longer-term goals for renewable power. The initial vehicle was the government's Long-Term Decarbonization Strategy (*Estrategia de Descarbonización a Largo Plazo* or ELP), developed in 2020. Building on the PNIEC's goal of 74% of electricity from renewable sources in 2030, it established objectives for 2040 and 2050 of 97% and 100%, respectively. Although the strategy did not set specific capacity targets, it noted that the further increase in installed capacity would not have to be as great as that required in the current decade, but also that completing the process of creating an entirely renewable electricity system would be more complex (MITECO 2020j).

The realization of these ambitious targets faced a number of challenges and obstacles, which can be summarized in three broad questions. First, could the desired amount of renewable capacity actually be built within the given time frame? Second, could so much new capacity be connected to and accommodated by the grid? Third, could the power generated meet demand, given the intermittent nature of so much of it? How the government planned to address each question is the focus of a separate section below.

## Can Sufficient Capacity Be Built?

The first challenge faced by the PNIEC was whether the desired amount of new renewable capacity could actually be built. According to the plan, the achievement of the targets would depend almost exclusively on private investment. Would the private sector in fact be willing to put up the required funds? Ultimately, this challenge was meant to be addressed by the establishment of yet another economic support scheme, with guarantees of future revenues awarded through a new and much more extensive series of auctions.

### *Preliminary Steps: Restoring Investor Confidence*

There were reasons to be concerned, dating back to the reforms introduced by the Rajoy government in 2013 and 2014 in order to reduce the tariff deficit. As a result of those drastic changes, as discussed in Chapter 4, owners of renewable installations saw significant cuts in their revenues, pushing some to the brink of bankruptcy and triggering a host of lawsuits. The government was strongly criticized for changing the rules retroactively, and many observers commented that irreparable damage had been done to investor confidence.

Those fears were belied to some extent by the subsequent willingness demonstrated by investors to build the capacity awarded in the auctions of 2016 and 2017, but the first Sánchez government nevertheless worried that more might need to be done to restore fully confidence in the stability of Spain's regulatory framework. As it happened, an important opportunity to do so presented itself in 2018 and 2019, as the deadline for updating the rate of return central to the Rajoy government reforms approached. The rates that had been established for the first six-year period—7.398% for plants already operating when the reforms went into effect—were set to expire at the end of 2019. Some observers feared that if no action were taken by the government, the rate of return could fall to as low as 4.5%—the current value of the 10-year government bond plus 300 basis points—for the 2020–2025 regulatory period. Overall, this would mean a further 25% cut in plant revenues, on top of the 40% that had already been absorbed (El Periódico de la Energía 2019a).

The Sánchez government began work on the issue almost as soon as it took office in 2018, asking the relevant regulatory agency, the National Commission on Markets and Competition (*Comisión Nacional de los*

*Mercados y la Competencia* or CNMC) to develop a recommendation. The CNMC in turn proposed a new methodology for determining the rate of return that, unlike the formula introduced in 2013, was commonly used by European regulators. Based on that recommendation, the government issued a draft law that would maintain the same rate of return (7.398%) for older installations for the next two 6-year periods, by which time many would be fully amortized. The government would, however, deduct any damages awarded in lawsuits from any compensation that was owed.

Consideration of the proposed law stalled as the parliament was dissolved and Spain went through the two national elections of 2019. Finally, with the deadline fast approaching, the caretaker Sánchez government resorted to the use of a decree-law (*Real Decreto-ley*), which would be subject to a simple up or down vote in the new parliament, in late November 2019. And given the stakes, even the People's Party, now the leader of the opposition, was reluctant to cast a negative vote. As with the draft law, the measure offered to maintain the existing rate of return for 12 years for pre-2013 producers who agreed to eschew legal action and to forfeit any awards. Otherwise, they would receive the rate based on the new methodology, 7.09%, which would be subject to further revision in six years. In this way, the government hoped to restore investor confidence as well as to put an end to the litigation, which already involved claims of some 10 billion euros, at an annual cost estimated to be at most 1500 million euros (Real Decreto-ley 17/2019).

*Market Forces*

Of course, even with faith in the stability of the regulatory system restored, the amount of investment in new renewable installations would still depend on how much money was to be made, and, in this regard, market forces offered a decidedly mixed message. On the one hand, construction and equipment costs had come down substantially, as discussed in Chapter 5, especially for PV, and they were expected to continue to decline. On the other hand, the introduction of increasing volumes of renewable power was likely to depress the wholesale price of electricity, reducing future revenue.

On the cost side of the equation, a 2020 study by the International Renewable Energy Agency (IRENA) found that the cost of utility-scale solar PV had fallen 82%, that of CSP by 47%, and that of onshore wind by

39% since 2010 (IRENA 2020), and a number of future-looking analyses predicted or assumed those trends would continue. For example, a 2019 International Energy Agency (IEA) world energy model used as inputs a further decline in the cost of solar PV of 40%, of onshore wind of 10%, and of CSP of more than 50% between 2018 and 2040 (IEA 2019). And a subsequent IRENA analysis found that the weighted average of the cost of renewable electricity in the Group of 20 nations could decline through the 2020s by up to 45% for onshore wind, 55% for PV, and 62% for CSP (Taylor 2020).

Nevertheless, a substantial amount of investment, more than 90 billion euros, was going to be required if the 2030 PNIEC targets for renewable capacity were to be met. And there were reasons to fear that the future wholesale price of electricity alone would not be sufficient to cover financing costs. The basic problem lay in the way the electricity market worked. Wholesale prices were determined by the merit order effect. As demand rises, generating sources with increasingly high marginal costs are summoned until the market clears, with the source with the highest marginal cost setting the market price. Some renewable sources, however, especially PV and wind, generate power at little or no marginal cost, and in the Spanish system, they have enjoyed priority dispatch, that is, the right to be called upon first. Thus the introduction of increasing amounts of wind and PV would tend to drive down the wholesale price, other things being equal, and as the price went down, financing for new capacity would tend to dry up, a process known by the colorful name of "cannibalization." According to one representative analysis, the wholesale price of electricity in 2030 was likely to be less than 10 euros per megawatt-hour (MWh) some 27% of the time, and possibly as much as 43% of the time, depending on the degree to which stabilizing measures were taken (Pöyry 2019).[1]

Investment in self-consumption installations was likely to be impacted particularly hard by falling prices. Because of their smaller size on average, they would tend to cost more per unit of installed capacity, with a levelized cost that was more likely to exceed the wholesale price. Thus it would not be profitable to invest in more generation than could be self-consumed, which would mean capping capacity at no more than 20–30% of maximum consumption, depending on the application (Revuelta 2020).

Early confirmation of such concerns arrived in late 2019, just as much of the capacity awarded in the auctions of 2016 and 2017 began to enter operation. For a couple of days in mid-December, the average pool price hovered around just 2 euros/MWh as renewables met more than 60%, and at moments nearly 80%, of demand (Roca 2019b, 2019c). The drop in demand occasioned by the COVID-19 pandemic had a similar effect. In March 2020, wholesale prices were down some 43% from the previous year, reaching as low as 9 euros/MWh (El Periódico de la Energía 2020a), while in Germany they briefly dropped into negative territory the following month (Amelang 2020).

### *The Solution: New Auctions*

The Sánchez governments nevertheless anticipated this problem. The PNIEC contained dozens of measures for promoting renewable energy and several specifically aimed at the development of new electric power installations. The most important of these measures was the use of auctions for the assignment of rights to specific levels of compensation for an extended period. In this way, the government could eliminate the price risk that would otherwise inhibit investment. The new auction scheme, however, would be much simpler than its predecessor. Awards would depend only on the price for electricity produced that was offered by bidders. Nevertheless, auctions could distinguish between different technologies based on their manageability, location, technical characteristics, and other criteria.

The government could not unilaterally introduce the new auctions, however. The pre-existing scheme was grounded in the Electricity Sector Law of 2013, and the law would have to be amended before the new auctions could proceed. The mechanism chosen for this purpose was the Sánchez government's proposed Law on Climate Change and the Energy Transition (*Ley de Cambio Climático y Transición Energética* or LCCTE).

In actuality, the history of the LCCTE long predated that of the PNIEC. The Congress had called for such a law as early as 2011, but there was no sense of urgency to develop one until after the Paris climate agreement in late 2015. Even then, the Rajoy government moved only slowly, in part because authority over the subject was divided between two ministries with different priorities. As a result, the conservative government could produce only an incomplete draft by the time it was pushed out of office in mid-2018.

The new Minister for the Ecological Transition, Teresa Ribera, whose portfolio included both energy and climate change for the first time, announced that the Sánchez government would present a new draft in September. Before it could do so, however, the Podemos party put forward its own detailed proposal containing a national renewable energy plan that placed considerable emphasis on the use of auctions in the event that market forces were not sufficient to achieve Spain's objectives. In the end, the Sánchez government did not circulate an informal draft until November 2018 and then issued a formal draft for public review the following February at the same time that it released the PNIEC. Notably, in addition to elaborating on the provisions already outlined in the PNIEC, the draft LCCTE set a very specific goal, subject to revision, of awarding rights to at least 3000 MW of renewable capacity each year, or enough to achieve roughly half of the targets contained in the PNIEC over a decade (MITECO 2019b).

Like the PNIEC, however, the draft law languished for many months as elections were called and the parties struggled to form a new government. During this time, the caretaker Sánchez government came under mounting pressure to hold another round of auctions, even if it meant using the Rajoy mechanism. Two years had passed since the last auctions, and further equipment orders were regarded as necessary to maintain Spain's newly revived production capacity. The caretaker government considered adding provisions to *Real Decreto-ley* 17/2019 that would enable it to hold new auctions as early as January, but it decided against doing so in the end so as not to jeopardize approval of the new rates of return for existing plants.

The coalition government formed in early 2020 issued a second formal draft of the LCCTE in February, which was reviewed by the CNMC, and submitted a final version for consideration by the Congress in May (MITECO 2020b). By then, the government had removed the auction target of at least 3000 MW per year in order to have more flexibility. The updates also allowed the product to be auctioned to be either power generated, production capacity, or both. And perhaps most importantly, the relevant provisions were reframed as formal amendments to the existing Electricity Sector Law, at the suggestion of the CNMC. The amendments would authorize the government to establish a new support scheme for renewable power as a replacement for the specific remunerative regime created by the Rajoy government.

Nevertheless, questions remained regarding how long it would take for the law to be ratified, certainly a matter of months, and early in the legislative process, the recently empowered right-wing opposition party,

*Vox*, presented an amendment to the entire proposal, which threatened to delay final approval until 2021. Given the urgency of achieving Spain's emissions reduction targets as well as the new imperative to promote economic recovery and job creation through investment in response to the coronavirus pandemic, the government decided not to wait. In late June, it approved a new decree-law, *Real Decreto-ley* 23/2020, that incorporated the auction provisions contained in the draft LCCTE and that would go into effect immediately, subject to a final up or down vote in the Congress within a month. And just days later, it issued a draft of the necessary implementing regulation (MITECO 2020c). As a result of this shift of focus, the first auctions were expected to take place during the second half of the year.

The new regulation was justified primarily in terms of market failure, specifically, the expected decline in prices due to the introduction of more renewable power. The resulting uncertainty over revenues would in turn make it difficult to obtain financing for new projects, thereby jeopardizing the attainment of Spain's renewable energy and decarbonization targets. To address this threat, the draft established a new framework for compensation, the Renewable Energy Economic Regime (*Régimen Económico de Energías Renovables* or REER). A guaranteed fixed price per unit of electricity would be paid to the lowest bidders in a series of auctions, where rights would be awarded on the basis of either power generation or capacity (or both). Generators would sell directly to the wholesale market. Any difference between the market price and the guaranteed price would be recovered not in access tariffs, as with the previous regimes, but as charges to marketers and direct consumers, who in turn would receive any savings when market prices were higher. In order to encourage production during times of peak demand, the government could award an incentive of up to 50% of the base price, while those who delivered less than the amount of power promised could be subject to penalties. Nevertheless, much additional detail, including a schedule of auctions, would need to be specified in a Ministerial Order and then a Resolution by the Secretary of State for Energy before the first auction could take place.

The draft regulation was subject to a barrage of criticism by the marketers, the major utilities, and even the CNMC. The proposed clearing mechanism would create major headaches for the marketers, whose costs would vary in unpredictable ways depending on the wholesale price and how much guaranteed electricity was offered at any given moment. Critics also argued that the system would distort price signals, encouraging consumption when prices were high and disincentivizing storage and even electrification. Some worried about the potential for market manipulation. And some simply maintained that auctions were not needed to meet the PNIEC targets; they regarded PV in particular as sufficiently mature to compete directly with other power sources (Monforte 2020a, 2020b; CNMC 2020a; AleaSoft Energy Forecasting 2020; Colón 2020).

In early November, the Council of Ministers approved the final draft as *Real Decreto* 960/2020. It contained few changes from the original draft, notwithstanding the intensity of the criticism. Then, a few days later, the newly retitled Ministry for the Ecological Transition and Demographic Challenge (*Ministerio para la Transición Ecológica y el Reto Demográphico* or MITECO) issued a draft of the required Ministerial Order, which laid out more detailed procedures and parameters of future auctions (MITECO 2020d). Initially, the product auctioned would be limited to installed capacity (not power generated). Bidders would have to deposit a guarantee of 60 euros per kilowatt (kW), separate from the 40 euros per kW required for access permits, and then meet a series of milestones or risk losing part or all of the guarantee. The document set both a cap and a floor for the number of hours of generation per year that would be compensated for each technology and a market participation incentive of either five or 25%, depending on the technology and whether the output of the installation could be managed.

The draft Ministerial Order also established the initial auction schedule, covering the 2020–2025 period. It called for auctioning at least 3100 MW of capacity in 2020, including a minimum of 1000 MW of both PV and wind. Over the entire six-year period, the government would auction rights to at least 10,000 MW of PV, 8500 MW of wind, 500 MW of CSP, 380 MW of biomass, and 60 MW of other technologies. If these goals were met, Spain would achieve a high percentage of the 2025 interim targets for PV and wind contained in the PNIEC.

The final Order (MITECO 2020f) was approved by the government in early December. It made minor changes in the schedule for CSP and biomass as well as the covered operating hours for PV and wind, and it slightly reduced the size of the first auction, to 3000 MW. Otherwise, however, it was largely unchanged from the first draft. The Order was followed a week later by the required Resolution of the Secretary of State for Energy, which set the date for the first auction as January 26,

Table 7.2 Auction schedule (minimum amounts of capacity in MW)

|  | 2020 | 2021 | 2022 | 2023 | 2024 | 2025 | Total by technology |
|---|---|---|---|---|---|---|---|
| Wind | 1000 | 1500 | 1500 | 1500 | 1500 | 1500 | 8500 |
| Solar PV | 1000 | 1800 | 1800 | 1800 | 1800 | 1800 | 10,000 |
| CSP | 0 | 200 | 0 | 200 | 0 | 200 | 600 |
| Biomass | 0 | 140 | 0 | 120 | 0 | 120 | 380 |
| Other technologies | 0 | 20 | 0 | 20 | 0 | 20 | 60 |
| Total by year | 2000 | 3640 | 3300 | 3620 | 3300 | 3620 | 19,100 |

*Source* MITECO (2020f, 111369)

2021, and fixed how long a given installation could receive remuneration under the new regime at 12–15 years, depending on the technology. Although Spain would not meet the government's goal of holding the first auction before the end of calendar year 2020, the country was poised to begin what could be the largest buildup of renewable power in its history (MITECO 2020g) (Table 7.2).

Indeed, the first auction was deemed a major success. The government received bids from some 84 companies totaling 9700 MW, or more than three times the target capacity. Of the total of 3034 MW awarded rights to a fixed price, 2036 MW went to PV projects and 998 MW to wind. The weighted average price of the winning bids came to 24.47 euros per MWh for PV and 25.31 euros per MWh for wind, or more than 40% below long-term wholesale electricity price forecasts at the time, with the lowest offer coming in at less than 15 euros per MWh. The resulting projects were expected to mobilize more than 2100 million euros in investment and create some 27,000 jobs and, once in operation, to bring down the market price by 1.3 euros per MWh. In order to continue to qualify for the fixed price, however, PV installations would have to be completed within two years and wind plants within three (MITECO 2021a).

## Can so Much Capacity Be Accommodated by the Electrical Grids?

It would not be enough simply to build all the renewable power capacity called for in the PNIEC, however. The electricity it generated would be of little use if it could not be connected to and accommodated by the electrical grid. And in both respects, there was at least some reason for concern.

### *Overall Grid Capacity*

Even before the rush to complete the installations awarded in the auctions of 2016 and 2017, renewable projects occasionally encountered insufficient grid capacity at the locations they desired to be built (Roca 2018). And the major electrical utilities estimated that investments in the grid of between 14 and 15 billion euros between 2017 and 2030 would be needed to integrate efficiently new self-consumption alone (Monforte 2019b).

Nevertheless, overall grid capacity was not expected to be a major problem. The most recent national planning document, for the six-year period 2015–2020, had called for only 4554 million euros of new investment in the transportation grid (MINETUR 2015). But in early 2019, the national grid operator, Red Eléctrica de España (REE), announced that it would invest 3 billion euros in 2018–2022 just to prepare the grid for renewables (Rojo Martín 2019).

The PNIEC itself called for investing 58.6 billion euros in grids and electrification, including 9.0 billion euros in the transportation network and international interconnections and 22.7 billion euros in distribution networks, largely to accommodate the planned increase in renewable generating capacity. The total amount for grids represented a more than 50% increase over previous expectations of what would be spent in these areas (MITECO 2020k, 24). At about the same time (March 2019), REE began work on the 2021–2026 planning document for the transportation grid, and the first draft issued early the following year contained nearly 6 billion euros, which would go a long way toward meeting the 10-year targets in the PNIEC (CNMC 2020d). In June 2020, REE confirmed that it had the financial capacity to meet the challenge posed by the PNIEC (El Periódico de la Energía 2020b).

One measure that would help minimize the additional investments required by optimizing the use of the existing grid was known as hybridization. This was where a developer could add storage or new generating capacity of a different renewable technology to an existing plant and use the same connection point, up to the previously approved volume. Hybrid plants were called for in the PNIEC, and the legal and regulatory foundations for them were laid in the LCCTE, *Real Decreto-ley* 23/2020, *Real Decreto* 960/2020, and a planned new regulation on access and connection (see below).

Another development that might make it easier for the grid to accommodate the growth in renewables was the freeing up of connection capacity by the planned closure of coal-fired and nuclear power plants. Under the PNIEC, all of the former and more than half of the latter were scheduled to be phased out by 2030. Including coal plant shutdowns since 2015, more than 15 GW of connection capacity would become available. In late 2019, the government modified the Electricity Sector Law so as to enable it to regulate the procedures and establish requirements for the concession of those rights to new renewable power installations (Real Decreto-ley 17/2019).

*Access and Connection*

Of greater concern, then, was the system for allocating permits to access and to connect to the grid. For a while, it appeared that many otherwise viable renewable power projects would have trouble acquiring the necessary permits. The trouble had its origins in the laissez-faire nature of the system. Virtually anyone could apply for access permits by putting down a guarantee of just 10 euros per kW of planned capacity. There was no need to take any further steps toward actually building a power plant that would generate electricity, and for permits awarded after 2013, there was no expiration date.

As the prospects for renewables began to improve during the run-up to the December 2019 deadline for the completion of the projects awarded in the auctions, this arrangement led to a burst of applications. At the end of January 2019, a total of 43 GW of new PV and wind capacity had been granted access permits, far more than the less than 9 GW that had been awarded in all the auctions, and another 53 GW in permit applications remained to be processed. By the end of the year, the number of applications had grown by more than 100 GW (REE n.d.).

Observers feared that many of these applications were primarily speculative in nature. Many of those receiving permits, it seemed, had no intention of building new renewable power plants themselves. Rather, they hoped to sell the permits to legitimate developers, possibly at a steep markup, which would effectively reduce the attractiveness of investment in PV and wind. According to some estimates, fewer than half of the awarded permits were associated with actual projects, and the cost of purchasing access permits reached as high as 250,000 euros per MW, potentially raising project costs by as much as 30% (El Periódico de la Energía 2019b; Ojea 2019a, 2020; Roca 2020a; Real Decreto-ley 23/2020).

A comprehensive updating of the increasingly outdated regulations on access and connection had been broadly called for since at least 2011. As one of its last acts, in May 2018, the Rajoy government issued a draft regulation, and the subsequent Sánchez government continued work on the subject. As a first step in seeking to quell speculative applications, the government quadrupled the required guarantee, from 10 to 40 euros per kW, in legal amendments adopted in the fall of 2018 (Real Decreto-ley 15/2018). But the new government's efforts to produce a complete draft regulation of its own was interrupted by the call for new national elections in early 2019.

While work was suspended at the level of the central administration, important developments occurred in other venues. In May 2019, REE began to deny applications on the grounds that there was insufficient grid capacity at the requested connection point or that no substation, necessary for connection, was planned at the proposed location, and over the remainder of the year, it denied more capacity in applications, more than 75 GW, than it approved. Then, in June 2019, the CNMC proposed guidance (*propuesta de Circular*) that sought to distinguish between applications for access permits that were backed by solid projects and speculative ones by linking the holding of the permits to the achievement of specific milestones, such as completing the required environmental review and obtaining authorization to begin construction (Ojea 2019b).

The CNMC proposal triggered a jurisdictional dispute between the Sánchez government and the regulatory body that ultimately had to be resolved, in the government's favor, by the Council of State, the country's supreme consultative body (Monforte 2020c). Nevertheless, the proposal's substance was reflected in the vehicle eventually chosen by the government to put an end to speculation. The first article of *Real Decreto-ley* 23/2020 established a timeline with five administrative milestones toward the completion of new renewable installations that existing and future permit holders would have to meet. In the event that a milestone was missed, the permit would expire and the developer would lose the 40 euro/kW guarantee. Permit holders who had no intention of meeting—or were not in a position to meet—the milestones received a grace period of three months to surrender the permits without losing the guarantee, and by the end of September, the holders of nearly 40 GW worth, which corresponded to some 22% of all permits under review or granted but not yet in service, had done so. *Real Decreto-ley* 23/2020 also established a moratorium on applications until new regulations were developed. Exceptions were made for self-consumption facilities and new plants taking advantage of the connection points freed by the closure of coal and nuclear plants (Roca 2020f).

This emergency measure did not eliminate speculation altogether. During the two days before *Real Decreto-ley* 23/2020 took effect, some 20 GW of access permit applications were made. And as the door closed temporarily to additional permits, some of the speculative activity shifted to the land where already-permitted projects might be built (Roca 2020b, 2020c).

Meanwhile, the government worked quickly, issuing a draft regulation in mid-July. According to this draft, the procedures for applying for access and connection would be combined into a single process. As a general rule, permits would be awarded in the order in which they were submitted, and applications would continue to have to be accompanied by a guarantee of 40 euros per kW. Nevertheless, exceptions were made to spur the penetration of renewable energy and electricity storage. These included applications involving newly hybridized installations (see above) and competitions limited to all new renewable installations or storage facilities for newly created, liberated, or expanded connection capacity on the transportation grid. In addition, the proposal simplified procedures for plants of up to 15 kW and exempted a range of installations intended primarily for self-consumption. The network manager would have up to 60 days to respond to applications, depending on the type of network (transportation versus distribution) and the size of the installation (MITECO 2020e).

In early September, the CNMC published its evaluation of the draft regulation, which was very positive. And a few days later, the regulatory body issued a revised draft of its own guidance on the methodology and conditions of access and connection, which was meant to complement the government's regulation (CNMC 2020b, 2020c). It was not until the very end of the year, however, that the government's Council of Ministers approved the final regulations (Real Decreto 1183/2020) governing access and connection. Hewing closely to the original draft, the regulation continued to promote renewable installations, hybridization, and storage. And it was not until January of 2021 that the CNMC issued the final version of its own circular concerning the more technical aspects of access and connection (CNMC 2021). The process of issuing permits could resume, albeit in a more deliberate and controlled manner than before.[2]

## Can the New Renewable Capacity Meet Demand?

A third general concern regarded the ability of renewable power to meet electricity demand, even if all the PNIEC targets were met. The problem here stemmed from the intermittent nature of much renewable generation, especially PV and wind. Even with nearly 90 GW of combined PV and wind capacity in 2030, there could be times when they alone would be unable to meet demand, especially given that overall electricity consumption was expected to grow with the increasing electrification of transportation and other parts of the economy (MITECO 2020a, 244).

Of course, PV and wind were not the only renewable sources in the mix. The PNIEC also called for some 32 GW of more manageable hydropower, CSP, and biomass. But even these might not suffice during periods of peak demand. Meanwhile, most conventional sources of dispatchable power—nuclear, co-generation, and coal—were scheduled to decline in capacity, with coal being phased out altogether. The one exception to this pattern was Spain's 26.6 GW of gas-fired combined cycle capacity (GFCC). Indeed, according to one unofficial estimate, Spain could need an additional 7 GW of GFCC to compensate for the loss of coal and nuclear power (Cepeda Minaya 2019). But that capacity could be hard to maintain, let alone increase, if the presence of so much renewable power drove down wholesale electricity prices sufficiently, and according to the PNIEC itself, GFCC plants would be used only 14% of the time on average, making it difficult if not impossible to stay afloat financially.

To address these concerns about balancing supply and demand after the integration of so much new renewable power, the PSOE-led administrations could—and in some cases did—adopt several types of measures. The most straightforward of these were capacity payments, intended to incentivize producers to maintain generating capacity that would otherwise be too expensive to operate. In 2007, as the amount of renewables took off, Spain introduced an investment subsidy for GFCC power plants that paid up to 10,000 euros per MW per year for 20 years. Along with certain other types of installations, GFCC was also eligible for an availability payment (*pago por disponibilidad*) of some 4500 euros per MW per year (Wynne and Julve 2016; MITECO 2020a, 282).

Although the new EU electricity market regulation published in 2019 prohibited such direct subsidies, it did provide for capacity mechanisms based on competitive processes, as long as they were temporary and a last resort (Regulation (EU) 2019/943; Monforte 2019a). With these conditions in mind, the Sánchez government opened a public consultation on the subject in September 2020. The reaction was mixed. Some proponents of renewables argued that Spain had no need for a capacity mechanism, while independent producers who had invested heavily in GFCC demanded the quick development of one (Roca 2020d, 2020e).

A second avenue would be to increase the amount of storage in the system, which in turn could take several forms. Spain already had 3.3 GW of pumped hydroelectric storage (*bombeo puro*) as well as thermal storage associated with the 2.3 GW of CSP. The PNIEC proposed to increase these amounts by 3.5 and 5.0 GW, respectively. The PNIEC also called for the introduction of 2.5 GW of battery storage, with a minimum of two hours of storage at maximum capacity (MITECO 2020a, 241). And relatively belatedly, the government recognized the potential value of hydrogen as a storage medium, one that could be used both to generate electricity and as a stand-alone fuel with applications in the transportation and industrial sectors. In addition to helping meet demand when renewable generation was low, storage could be used to minimize the electricity losses when renewable output exceeded consumption by absorbing the excess production.

The field of storage was a particularly fluid one, however, featuring rapid innovations that made it difficult to keep abreast of technological developments. Nevertheless, the government did what it could to create a legal and regulatory framework that would facilitate and promote the introduction of cost-effective storage. As a first step, the LCCTE and the derivative *Real Decreto-ley* 23/2020 formally recognized storage operators as legal subjects in the electrical system. Specific legal amendments were also included to facilitate the development of reversible hydroelectric stations as pumped storage systems.

Then, in a parallel process beginning in April 2020, the coalition government launched the development of a National Storage Strategy and a Hydrogen Roadmap, reflecting the rapid growth of interest in the latter in particular as a promising storage medium. By July, it had prepared a draft of the hydrogen roadmap that called for 4 GW of electrolytic conversion capacity by 2030 and that was approved in October. At that time, the government also issued a draft comprehensive energy storage strategy, which called for a total of 20 GW in 2030 and 30 GW in 2050, primarily to support the electrical system (MITECO 2020h, 2020i).

In February 2021, the government approved the final version of Energy Storage Strategy, which contained the same top-level goals. Spain would have to add nearly 12 new GW of storage by 2030 and another 10 GW in the following decade. The strategy envisioned key roles for pumped hydroelectric, batteries, thermal storage associated with CSP plants, and hydrogen as well as both large-scale and smaller distributed storage facilities (MITECO 2021b).

On the other side of the equation, steps could be taken to manage demand so that a balance was easier to maintain. Spain had recently established an interruptibility service whereby large consumers could agree, for a fee, to reduce their usage by pre-determined amounts when ordered by the grid operator. But there was much more the government could do to promote demand management more broadly, as recognized in the PNIEC (Roldán Fernández et al. 2016; MITECO 2020a, 67–69).

Once again, broad reforms in regulation, market organization, and business models would be necessary to take full advantage of the potential for demand management in the system. As a first step, the LCCTE and *Real Decreto-ley* 23/2020 proposed to create the legal subject of independent aggregator, who could play a distinct role in the electricity market by combining loads from diverse consumers. The PNIEC also called for demand management plans, pilot projects, and simplifying the administrative procedures associated with demand management (PNIEC 52, 87–88).

Should these various domestic measures still not suffice to integrate successfully so much renewable power, Spain could employ one further course of action: developing additional interconnections with the broader European electricity market through France. Because Spain, and the Iberian Peninsula more generally, is an isolated "energy island," interconnections that enable Spain to export and import electricity as needed are vital for balancing supply and demand. In the late 2010s, Spain doubled the amount of power it could export across its northern border, to nearly 2.8 GW, or about 6% of peak power consumption, and Spain and France are building an underwater interconnection in the Bay of Biscay with a capacity of 2.2 GW, to be completed by 2025 with the help of an

EU grant of nearly 600 million euros—the largest EU grant ever for energy infrastructure. Two additional crossings in the Pyrenees that could increase the total to 8 GW are expected to be operational by 2027. The ultimate goal is to bring Spain's total interconnection capacity, including 3 GW with Portugal, up to the EU's target for 2030 of 15% of Spain's electricity consumption (EC 2018; MITECO 2020a, 17, 32, 64–66).

## Conclusion

By early 2021, the PSOE-Podemos coalition government had established ambitious targets for renewable power and a broad plan for achieving those targets while ensuring the security of Spain's electricity supply. Central to this plan was an expected series of auctions that would stimulate investment in at least 20 GW of new renewable capacity over the next five years. But it also included numerous measures to ensure that the new capacity could be connected to the grid and that the additional power it generated could be seamlessly integrated into the electricity supply. Nevertheless, although Spain successfully held the first auction under the new economic regime in early 2021, it would be some time before one could assess how likely the achievement of Spain's renewable power targets would be and what adjustments in both ends and means might need to be made.

This chapter provides further evidence of the centrality of the core executive in shaping Spanish renewable power policy. All the major legislation, regulation, and plans were drafted in the new Ministry for the Ecological Transition and its subordinate agencies and adopted largely as is. Although a centerpiece of the government's program, the LCCTE, was delayed by the need to hold elections and then opposition in the Congress, the executive was deftly able to use the mechanism of the decree-law (*Real Decreto-ley*), intended for cases of "extraordinary and urgent need," to make desired legislative changes that the Congress had little choice but to approve.

In early 2019, the competencies of the CNMC were broadened in order to bring Spain into compliance with EU legislation (Real Decreto-ley 1/2019), and the regulatory body began to play a more central role. Nevertheless, when its activities appeared to challenge the authority of the core executive, the latter prevailed. Likewise, the expressed views of important interest groups seemed to have little influence. Although existing producers received an acceptable rate of return for the next regulatory period, the government had independent reasons for being relatively generous and, in the process, drove a hard bargain with regard to legal remedies. And when it came to devising the details of the new auction system, the many concerns expressed appeared to go largely unheeded.

## Notes

1. For an early discussion of cannibalization, see Sanchis and Fernández (2018).
2. In the meantime, the government had begun the process of developing regulations for the concession of the connection capacity freed by the closure of coal-fired and nuclear power plants (El Periódico de la Energía 2020c).

## References

AleaSoft Energy Forecasting. 2020. "Las subastas de renovables, con pinzas" (21 Oct.) https://aleasoft.com/es/subastas-renovables-con-pinzas/.

Amelang, Sören. 2020. "Negative Electricity Prices." *energypost.eu* (18 May). https://energypost.eu/negative-electricity-prices-lockdowns-demand-slump-exposes-inflexibility-of-german-power/.

Cepeda Minaya, Denisse. 2019. "Lo bueno y lo malo del apagón del carbón y la nuclear." *CincoDías* (26 Feb.). https://cincodias.elpais.com/cincodias/2019/02/25/companias/1551117531_788571.html.

Colón, Javier. 2020. "El decreto de subastas de renovables y la manipulación del mercado." *El Periódico de la Energía* (5 Nov.). https://elperiodicodelaenergia.com/el-decreto-de-subastas-de-renovables-y-la-manipulacion-del-mercado/.

CNMC (Comisión Nacional de los Mercados y la Competencia). 2020a. "Acuerdo por el que Se Emite Informe sobre el Proyecto de Real Decreto por el que Se Regula el Régimen Económico de Energías Renovables para Instalaciones de Producción de Energía Eléctrica." IPN/CNMC/014/20 (30 July). https://www.cnmc.es/sites/default/files/3086911_0.pdf.

———. 2020b. "Acuerdo por el que Se Emite Informe sobre el Proyecto de Real Decreto de Acceso y Conexión a las Redes de Transporte y Distribución de Energía Eléctrica." IPN/CNMC/022/20 (2 Sept.). https://www.cnmc.es/sites/default/files/3139862.pdf.

———. 2020c. "Propuesta de Circular X/2020, de Xxx de Xxx, de la Comisión Nacional de los Mercados y la Competencia, por la que Se Establece la Metodología y Condiciones del Acceso y de la Conexión a las Redes de

Transporte y Distribución de las Instalaciones de Producción de Energía Eléctrica" (Sept.). https://www.cnmc.es/sites/default/files/editor_contenidos/Energia/Consulta%20Publica/01_Nueva_Propuesta_Circular_Acceso_y_Conexion_2020.pdf.

———. 2020d. *Planificación de la Red de Transporte de Energía Eléctrica: Período 2021–2026.* INF/DE/005/20 (3 Apr.). https://www.cnmc.es/expedientes/infde00520.

———. 2021. *Circular 1/2021, de 20 de enero, de la Comisión Nacional de los Mercados y la Competencia, por la que se establece la metodología y condiciones del acceso y de la conexión a las redes de transporte y distribución de las instalaciones de producción de energía eléctrica.* (22 Jan.). https://www.cnmc.es/sites/default/files/3332956_4.pdf.

Comisión (Comisión de Expertos de Transición Energética). 2018. *Análisis y Propuestas para la Descarbonización.* http://www6.mityc.es/aplicaciones/transicionenergetica/informe_cexpertos_20180402_veditado.pdf.

Directive (EU) 2018/2001 of the European Parliament and of the Council of 11 December 2018 on the promotion of the use of energy from renewable sources. *Official Journal of the European Union* (21 Dec. 2018). https://eur-lex.europa.eu/legal-content/EN/TXT/PDF/?uri=CELEX:32018L2001&from=EN.

EC (European Commission). 2018. "European Solidarity on Energy: Better Integration of the Iberian Peninsula into the EU Energy Market" (27 July). https://ec.europa.eu/commission/presscorner/detail/en/IP_18_4621.

———. 2019. "Assessment of the Draft National Energy and Climate Plan of Spain." Commission Staff Working Document SWD(2019) 2562 final (18 June.) https://ec.europa.eu/energy/sites/ener/files/documents/es_swd_en.pdf.

El Periódico de la Energía. 2019a. "Anpier avisa a Ribera: el sector fotovoltaico saltará por los aires si hay un nuevo recorte a las primas" (13 June). https://elperiodicodelaenergia.com/anpier-avisa-a-ribera-el-sector-fotovoltaico-saltara-por-los-aires-si-hay-un-nuevo-recorte-a-las-primas/.

———. 2019b. "La CNMC quiere poner coto a los abusos y a la especulación de los puntos de acceso y conexión a la red" (6 June). https://elperiodicodelaenergia.com/la-cnmc-quiere-poner-coto-a-los-abusos-y-a-la-especulacion-de-los-puntos-de-acceso-y-conexion-a-la-red/.

———. 2020a. "Altran alerta sobre la 'canibalización' de las renovables" (23 April). https://elperiodicodelaenergia.com/altran-alerta-sobre-la-canibalizacion-de-las-renovables-el-precio-estara-a-cero-muchas-horas-por-la-mayor-penetracion-de-solar-y-eolica/.

———. 2020b. "Beatriz Corredor defiende que Red Eléctrica está preparada para acometer las inversiones del PNIEC" (15 June). https://elperiodicodela

energia.com/beatriz-corredor-defiende-que-red-electrica-esta-preparada-para-acometer-las-inversiones-del-pniec/.

———. 2020c. "El Gobierno lanza a consulta pública el concurso para la concesión de los nudos de red afectados por los cierres de térmicas" (11 Dec.). https://elperiodicodelaenergia.com/el-gobierno-lanza-a-consulta-publica-el-concurso-para-la-concesion-de-los-nudos-de-red-afectados-por-los-cierres-de-termicas/.

European Council. n.d. "The 2030 Climate and Energy Framework." https://www.consilium.europa.eu/en/policies/climate-change/2030-climate-and-energy-framework/.

IEA (International Energy Agency). 2019. "World Energy Model" (Oct.). https://www.iea.org/reports/world-energy-model/techno-economic-inputs#abstract.

IRENA (International Renewable Energy Agency). 2020. *Renewable Power Generation Costs in 2019.* https://www.irena.org/-/media/Files/IRENA/Agency/Publication/2020/Jun/IRENA_Power_Generation_Costs_2019.pdf.

MINETUR (Ministerio de Industria, Energía y Turismo). 2015. *Planificatión Energética: Plan de Desarrollo de la Red de Transporte de Energía 2015–2020.* https://www.mincotur.gob.es/energia/planificacion/Planificacionelectricidadygas/desarrollo2015-2020/Documents/Planificaci%C3%B3n%202015_2020%20%202016_11_28%20VPublicaci%C3%B3n.pdf.

MITECO (Ministerio para la Transición Ecológica). 2019a. "Borrador del Plan Nacional Integrado de Energía y Clima 2021–2030" (Feb.). https://www.miteco.gob.es/images/es/documentoparticipacionpublicaborradordelplannacionalintegradodeenergiayclima2021-2030_tcm30-487344.pdf.

———. 2019b. "Anteproyecto de Ley de Cambio Climático y Transición Energética" (20 Feb.). https://www.miteco.gob.es/images/es/1anteproyectoleyccyte_tcm30-487336.pdf.

MITECO (Ministerio para la Transición Ecológica y Reto Demográfico). 2020a. "Borrador Actualizado del Plan Nacional Integrado de Energía y Clima 2021–2030" (20 Jan.). https://www.miteco.gob.es/images/es/pnieccompleto_tcm30-508410.pdf.

———. 2020b. "Proyecto de Ley de Cambio Climático y Transición Energética" (19 May). https://www.miteco.gob.es/es/ministerio/proyecto-de-ley-de-cambio-climatico-y-transicion-energetica.aspx.

———. 2020c. "Proyecto de Real Decreto por el que Se Regula el Régimen Económico de Energías Renovables para Instalaciones de Producción de Energía Eléctrica" (25 June). https://energia.gob.es/es-es/Participacion/Paginas/DetalleParticipacionPublica.aspx?k=324.

———. 2020d. "Orden por la que Se Regula el Primer Mecanismo de Subasta para el Otorgamiento del Régimen Económico de Energías Renovables y Se Establece el Calendario Indicativo para el Periodo 2020–2025" (6 Nov.). https://energia.gob.es/es-es/Participacion/Paginas/Detall eParticipacionPublica.aspx?k=356.

———. 2020e. "Propuesta de Real Decreto de acceso y conexión a las redes de transporte y distribución" (June). https://energia.gob.es/es-ES/Participa cion/Paginas/Cerradas/RD-acceso-conexion-redes-transporte.aspx.

———. 2020f. "Orden TED/1161/2020, de 4 de diciembre, por la que se regula el primer mecanismo de subasta para el otorgamiento del régimen económico de energías renovables y se establece el calendario indicativo para el periodo 2020–2025" (4 Dec.). https://www.boe.es/boe/dias/2020/12/05/pdfs/BOE-A-2020-15689.pdf.

———. 2020g. "Resolución de 10 de diciembre de 2020, de la Secretaría de Estado de Energía, por la que se convoca la primera subasta para el otorgamiento del régimen económico de energías renovables al amparo de lo dispuesto en la Orden TED/1161/2020, de 4 de diciembre" (12 Dec.). https://www.boe.es/boe/dias/2020/12/12/pdfs/BOE-A-2020-16068.pdf.

———. 2020h. *Hoja de Ruta de Hidrógeno: Una Apuesta por el Hidrógeno Renovable* (Oct.). https://www.miteco.gob.es/es/ministerio/hoja_de_ruta_del_hidrogeno_una_apuesta_por_el_hidrogeno_renovable_tcm30-513830.pdf.

———. 2020i. "Borrador de da Estrategia de Almacenamiento Energético" (Oct.). https://elperiodicodelaenergia.com/wp-content/uploads/2020/10/Borrador-de-la-Estrategia-de-Almacenamiento.pdf.

———. 2020j. *Estrategia de Descarbonización a Largo Plazo* (Nov.). https://www.miteco.gob.es/es/prensa/documentoelp_tcm30-516109.pdf.

———. 2020k. "Impacto Económico, de Empleo, Social y sobre la Salud Pública del Borrador Actualizado del Plan Nacional Integrado de Energía y Clima 2021–2030" (20 Jan.). https://www.miteco.gob.es/images/es/pniec_2021-2030_informesocioeconomico_borradoractualizado_tcm30-506495.pdf.

———. 2021a. "El MITECO celebra la primera subasta renovable del periodo 2020–2025 para facilitar la acción climática y reducir la factura eléctrica" (26 Jan.). https://www.miteco.gob.es/es/prensa/210126resultadosprimeras ubastaderenovablesd_tcm30-522089.pdf.

———. 2021b. *Estrategia de Almacenamiento Energético* (Feb.). https://www.cnmc.es/sites/default/files/3332956_4.pdf.

Monforte, Carmen. 2019a. "Las eléctricas estiman 15.000 millones de inversión para impulsar el autoconsumo." *CincoDías* (17 April). https://cincodias.elp ais.com/cincodias/2019/04/16/companias/1555437557_454461.html.

———. 2019b. "Eléctricas e industrias pujarán en las mismas subastas por el incentivo de capacidad." *CincoDías* (21 June). https://cincodias.elpais.com/cincodias/2019/06/20/companias/1561054065_140535.html.

———. 2020a. "La regulación de las subastas de renovables pone en pie de guerra al sector eléctrico." *CincoDías* (13 July). https://cincodias.elpais.com/cincodias/2020/07/02/companias/1593712531_799008.html.

———. 2020b. "Las grandes eléctricas critican el sistema con que se liquidará a las renovables." *CincoDías* (27 July). https://cincodias.elpais.com/cincodias/2020/07/26/companias/1595794962_615326.html.

———. 2020c. "La CNMC sale derrotada frente a Ribera en la pugna por las conexiones eléctricas." *CincoDías* (2 July). https://cincodias.elpais.com/cincodias/2020/07/02/companias/1593669707_740312.html.

Ojea, Laura. 2019a. "Las renovables no se fían del todo de las distribuidoras: prefieren solicitar el acceso a la red a REE." *El Periódico de la Energía* (6 Dec.). https://elperiodicodelaenergia.com/las-renovables-no-se-fian-del-todo-de-las-distribuidoras-prefieren-solicitar-el-acceso-a-la-red-a-ree/.

———. 2019b. "Las renovables no se fían del todo de las distribuidoras: prefieren solicitar el acceso a la red a REE." *El Periódico de la Energía* (12 June). https://elperiodicodelaenergia.com/las-renovables-no-se-fian-del-todo-de-las-distribuidoras-prefieren-solicitar-el-acceso-a-la-red-a-ree/.

———. 2020. "La COVID-19 acaba con la especulación de los puntos de acceso y conexión para los nuevos proyectos de renovables." *El Periódico de la Energía* (21 April). https://elperiodicodelaenergia.com/la-covid-19-acaba-con-la-especulacion-de-los-puntos-de-acceso-y-conexion-para-los-nuevos-proyectos-de-renovables/.

Pöyry. 2019. "Spain's National Energy and Climate Plan 2030: Madrid Energy Breakfast" (5 April).

Real Decreto 960/2020, de 3 de noviembre, por el que se regula el régimen económico de energías renovables para instalaciones de producción de energía eléctrica (4 Nov. 2020). https://www.boe.es/eli/es/rd/2020/11/03/960.

Real Decreto 1183/2020, de 29 de diciembre, de acceso y conexión a las redes de transporte y distribución de energía eléctrica (30 Dec. 2020). https://www.boe.es/eli/es/rd/2020/12/29/1183/con.

Real Decreto-ley 17/2019, de 22 de noviembre, por el que se adoptan medidas urgentes para la necesaria adaptación de parámetros retributivos que afectan al sistema eléctrico y por el que se da respuesta al proceso de cese de actividad de centrales térmicas de generación (23 Nov. 2019). https://www.boe.es/eli/es/rdl/2019/11/22/17.

Real Decreto-ley 23/2020, de 23 de junio, por el que se aprueban medidas en materia de energía y en otros ámbitos para la reactivación económica (24 June 2020). https://www.boe.es/eli/es/rdl/2020/06/23/23/con.

REE (Red Eléctrica de España). n.d. "Evolución de la tramitación de los procedimientos de acceso a la red de la generación eólica y solar fotovoltaica gestionados por Red Eléctrica" (various months). https://www.ree.es/es/clientes/datos-acumulados-generacion-renovable

Regulation (EU) 2018/1999 Of the European Parliament and of the Council of 11 December 2018 on the Governance of the Energy Union and Climate Action. *Official Journal of the European Union* (21 Dec. 2018). https://eur-lex.europa.eu/legal-content/EN/TXT/PDF/?uri=CELEX:32018R1999&from=EN.

Regulation (EU) 2019/943 of the European Parliament and of the Council of 5 June 2019 on the Internal Market for Electricity. *Official Journal of the European Union* (14 June 2019). https://eur-lex.europa.eu/legal-content/EN/TXT/PDF/?uri=CELEX:32019R0943&from=EN.

Revuelta, Javier. 2020. "How to Capture the Spanish Sun." *AFRY Insights* (Jan.). https://afry.com/sites/default/files/2020-01/afryinsights_jan2020.pdf.

Roca, Ramón. 2018. "Atasco en las redes eléctricas: una empresa se queda sin hacer una planta fotovoltaica de 74,5 MW por exceso de potencia en Badajoz." *El Periódico de la Energía* (16 Oct.). https://elperiodicodelaenergia.com/atasco-en-las-redes-electricas-una-empresa-se-queda-sin-hacer-una-planta-fotovoltaica-de-745-mw-por-exceso-de-potencia-en-badajoz/.

———. 2019a. "El Gobierno presentará «en los próximos días» una nueva versión del Plan Nacional de Energía y Clima." *El Periódico de la Energía* (24 Sept.). https://elperiodicodelaenergia.com/el-gobierno-presentara-en-los-proximos-dias-una-nueva-version-del-plan-nacional-de-energia-y-clima/.

———. 2019b. "Supersábado Eléctrico: el pool a 2,5 €/MWh, las renovables cubrirán el 80%, la nuclear reduce potencia y sin carbón y casi sin gas en el mix." *El Periódico de la Energía* (21 Dec.). https://elperiodicodelaenergia.com/supersabado-electrico-el-pool-a-25-e-mwh-las-renovables-cubriran-el-80-la-nuclear-reduce-potencia-y-sin-carbon-y-casi-sin-gas-en-el-mix/.

———. 2019c. "Fin de semana de records: el día que el 'pool' estuvo a punto de volver a marcar cero €/MWh." *El Periódico de la Energía* (23 Dec.). https://elperiodicodelaenergia.com/fin-de-semana-de-records-el-dia-que-el-pool-estuvo-a-punto-de-volver-a-marcar-cero-e-mwh/.

———. 2020a. "Endesa denuncia la alta especulación por los puntos de conexión: «Se pagan hasta 250.000 €/MW»." *El Periódico de la Energía* (20 Feb.). https://elperiodicodelaenergia.com/endesa-denuncia-la-alta-especulacion-por-los-puntos-de-conexion-se-pagan-hasta-250-000-e-mw/.

———. 2020b. "La especulación con las renovables no se corta." *El Periódico de la Energía* (16 July). https://elperiodicodelaenergia.com/la-especulacion-con-las-renovables-no-se-corta-el-nuevo-rdl-traslada-la-burbuja-al-suelo-con-precios-que-se-han-duplicado-en-cuestion-de-semanas/.

———. 2020c. "La ley antiespeculativa de Ribera provoca un aluvión récord de solicitudes de acceso de renovables en 24 horas." *El Periódico de la Energía* (31 July). https://elperiodicodelaenergia.com/la-ley-antiespeculativa-de-ribera-provoca-un-aluvion-record-de-solicitudes-de-acceso-de-renovables-en-24-horas/.

———. 2020d. "España no necesita mecanismos de capacidad para conseguir un suministro fiable." *El Periódico de la Energía* (21 Sept.). https://elperiodicodelaenergia.com/espana-no-necesita-mecanismos-de-capacidad-para-conseguir-un-suministro-fiable/.

———. 2020e. "Los productores independientes de energía eléctrica exigen al Gobierno un mercado de capacidad para hacer viables los ciclos combinados." *El Periódico de la Energía* (1 Oct.). https://elperiodicodelaenergia.com/los-productores-independientes-de-energia-electrica-exigen-al-gobierno-un-mercado-de-capacidad-para-hacer-viables-los-ciclos-combinados/.

———. 2020f. "El Gobierno consigue expulsar a casi 40 GW de proyectos especuladores en renovables y devuelve la friolera de 1.572 millones en avales." *El Periódico de la Energía* (3 Dec.). https://elperiodicodelaenergia.com/el-gobierno-consigue-expulsar-a-casi-40-gw-de-proyectos-especuladores-en-renovables-y-devuelve-la-friolera-de-1-572-millones-en-avales/.

Rojo Martín, José. 2019. "Spanish Grid Set for Funding Boom to Accommodate Renewables." *PV-Tech* (22 March). https://www.pv-tech.org/news/spanish-grid-set-for-funding-boom-to-accommodate-renewables.

Roldán Fernández, J.M., M. Burgos Payán, J.M. Riquelme Santos, J.M., and A.L Trigo García. 2016. "Renewable Generation Versus Demand-Side Management. A Comparison for the Spanish Market." *Energy Policy* 96 (Sept.): 458–470.

Sanchis, Antonio, and Alejandro Fernández. 2018. "El factor de apuntamiento y el efecto de canibalización en la fotovoltaica. Una perspectiva de future" (27 Nov.). https://blog.altran.es/industria-energia/factor-de-apuntamiento-y-efecto-canibalizacion-fotovoltaica/.

Taylor, Michael. 2020. "Analysis Shows Wind and Solar Costs Will Continue to Fall Dramatically Throughout the 2020s." *energypost.eu* (6 Nov.). https://energypost.eu/analysis-shows-wind-and-solar-costs-will-continue-to-fall-dramatically-throughout-the-2020s/.

Wynne, Gerard, and Javier Julve. 2016. "Spain's Capacity Market: Energy Security or Subsidy?" (Dec.). https://ieefa.org/wp-content/uploads/2017/11/Spains-Capacity-Market-Energy-Security-or-Subsidy_December-2016.pdf.

# CHAPTER 8

# The Politics of Renewable Power in Spain

## INTRODUCTION

The previous six chapters have reviewed the evolution of renewable power policy in Spain since the first years of the current parliamentary monarchy. During these four decades, we have observed a pattern of variation in the level of support, which has in turn contributed to alternating periods of growth and stagnation in generating capacity. A gradual increase in the 1980s and 1990s was followed by rapid buildups in first wind and then solar power in the first decade of the 2000s, thanks in no small part to favorable developments in policy. Then the level of support was drastically reduced, resulting in little no growth in renewable capacity during much of the 2010s. Most recently, government policy has become much more supportive, and developers have responded with a good deal of wind and record amounts of new solar power, with the promise of a doubling of Spain's renewable capacity in the decade ahead.

The goal of this final chapter is to provide an overarching explanation of the pattern of policy outcomes observed. To this end, it seeks to identify the key actors in the policymaking process, their interests and sources of influence, and, ultimately, the contributions that each has made.

The principal findings tend to confirm the expectations in the existing literature on public policy in Spain, as discussed in Chapter 1. The dominant actor in this policy arena has been the central executive of the

Spanish government. The executive, led by the President and the relevant ministers, has been the leading source of policy initiatives, regardless of whether the intention has been to promote or to throttle the deployment of renewable generating capacity. Even where the executive has not enjoyed a dependable majority in the Congress, it has been able to pursue its interests via decree-laws, decrees, and lesser forms of regulation.

Less clear has been the source of the executive's policy preferences, that is, what it has wanted to do with its power and authority. And here we encounter another puzzle: Governments led by different parties have often pursued similar policies. Variation in policy outcomes cannot be simply explained in terms of which major party was in office.

Thus much of the chapter is devoted to identifying the sources of preferences that have been common to the major parties and that have pushed them in similar directions. It begins by examining the material circumstances in which Spanish governments have found themselves. This is followed by a look at the international social structures in which Spain has had to operate, primarily the European Union. At the domestic level, one can point to common national imperatives such as reducing foreign energy dependence, industrial development, and job creation as well as fiscal solvency.

Nevertheless, there have been noteworthy differences of emphasis as well as times when the major parties have openly clashed. To account for these, we must turn to differences in party ideology and receptiveness to interest groups as well as the personalities of key individuals.

Finally, the chapter also considers the roles and activities of other national actors and the regional and municipal governments. Many of these have been quite active in promoting renewable power. Because of their relatively limited authority and resources, however, their impact has been mainly on the margins.

## THE PRIMACY OF THE EXECUTIVE

The evolution of renewable power policy in Spain confirms the findings of the literature on Spanish public policy more generally. As discussed in Chapter 1, policy making is characterized by a high degree of power concentration, with the core executive playing the key role in the policy process. Indeed, the President is constitutionally empowered to monopolize the most important decisions over national policy (Heywood 1999).

In the case of policy regarding renewable power, the most important actor has been the central government or, perhaps more precisely, a few key positions within the executive, including the Presidency, the ministry holding the energy portfolio, and, within the relevant ministry, the office of the Secretary of State for Energy. Indeed, the previous chapters are replete with examples of actions by the central government that determined important policy outcomes. These include the establishment and refinement of the special regime, many of the subsequent reductions in support for renewable power, the holding of auctions in 2016 and 2017, the various measures first impeding and then facilitating investment in self-consumption facilities, the preparation of multiple plans promoting renewable power, and more.

To be sure, precisely how much power the executive is able to wield may depend on a variety of contingent factors, such as the internal unity of the governing party and the personal leadership capacity of the President (Heywood 1999, 106). In the case of renewable power policy, particularly important has been whether the governing party has enjoyed an effective majority in the Congress. For much of the history of democratic Spain, a single party has been able to command a majority, by itself or with the help of other parties that it could reliably count on to pass legislation.

Beginning with the second Zapatero government (2008–2011) and then with every election since 2015, however, new governments have been able to form only at the sufferance of one or more other parties that have merely abstained during the vote of investiture. As a result, it has been more difficult if not impossible to pass desired laws, and on occasion, the government has been reduced to blocking legislation initiated by others through its control of the Bureau or Presiding Committee of the Congress (*Mesa*), as exemplified by the second Rajoy government's response to the multi-party legislative proposal regarding self-consumption. Thus, on only two occasions has the Congress passed entirely new laws, the electricity sector laws of 1997 and 2013, that were needed to lay the groundwork for government initiatives.

Nevertheless, the core executive has been able to work around this obstacle through the careful employment of decree-laws (*Real Decreto-ley*), which typically amend existing legislation. The use of decree-laws is meant to be limited to cases of extraordinary and urgent need and, as a result, they cannot be amended and are subject only to an up or down vote in the Congress. Thus a sponsoring government must be careful to include only what can be justified as urgent and extraordinary, or risk the measure's rejection. Despite these conditions, every government since the second Zapatero administration has employed this tactic successfully on multiple occasions.

Yet other policy initiatives have merely required that the government issue appropriate regulations, which can be done without any Congressional involvement. Indeed, the previous chapters are full of examples of both majority and minority governments approving regulations (*Real Decreto*), issuing ministerial orders, and preparing energy plans. In practice, such measures have been the principal vehicles for establishing specific renewable energy targets and spelling out the details of the mechanisms for promoting renewable power. Although their issuance has depended on the existence of permissive legislation, especially the electricity sector laws of 1997 and 2013, any limitations in this regard have often been removed by the passage of an appropriately tailored decree-law.

## SOURCES OF CONVERGENT POLICY PREFERENCES

To what ends have successive governments sought to employ this power? Typically, as a first approximation, we may expect the policy objectives of the government to vary with the party or parties in power, but especially the leading party of any coalition government. A striking finding of the previous chapters, however, is that the policies of the two main parties, the center-right People's Party (*Partido Popular* or PP) and the center-left Spanish Socialist Workers' Party (*Partido Socialista Obrero Españo* or PSOE), have been remarkably similar over the years when it comes to the promotion of renewable power. Beginning in the late 1990s and continuing through the mid- to late 2000s, successive PP and PSOE governments introduced and sought to refine the original support scheme, the special regime, in order to promote the growth of renewable power. Then, beginning in the late 2000s and continuing through the mid-2010s, successive PSOE and PP governments sought to reign in the costs of the special regime. And beginning in the mid- to late 2010s and continuing into the early 2020s, successive PP and PSOE (or PSOE-led) governments have sought once again to increase Spain's renewable power capacity. These commonalities reflected the balance of imperatives facing whatever party was in power during a particular period.

What has accounted for these similarities in policy preferences? We can point to at least three sets of factors: material conditions, the international political environment, and domestic imperatives. For the most part, these factors have militated in favor of promoting renewable power, although the third set in particular has been at times a source of pressures to limit its growth as well.

### *Material Conditions*

Two material conditions stand out when attempting to understand the evolution of renewable power policy in Spain. One is Spain's prodigious renewable power potential. From the beginning, Spain's solar resources have been regarded as immense, as much as several terawatts, or much more than the country could ever use, thanks to Spain's extensive land area and the highest levels of solar radiation in Europe. Meanwhile, government estimates for Spain's wind power potential have marched steadily upward, from about 15 gigawatts (GW) in 1999 to more than 40 GW in 2005 and then to as many as 330 GW in 2011 (IDAE 1999, 58–59; 2005, 42, 161; 2011, xxxvi, 237, 379–381).

Nevertheless, it is challenging, if not impossible, to evaluate a country's renewable power potential independently of the technology available to exploit it. Indeed, as suggested by the steady increases in estimates for wind power above, forecasts of that potential can be tightly bound up with current assessments and future projections of the state of technology, which is critical for determining the cost-effectiveness of investment. Thus technology is a second important material condition. And in the case of renewable power, as we have seen in previous chapters, the principal theme has been that of a relatively steady decline in the costs of equipment due in large part to technical advances. By some estimates, the capital costs of solar PV declined by as much as 90% during the 2010s alone, and those for wind have come down over the years to a significant extent as well.

To some degree, this was the whole point of support for renewable power, especially in the first decade of the century: to drive down the costs. In that regard, technology was an endogenous factor. But given the relatively small size of the Spanish market, the declines observed almost certainly had more to do with larger global market forces over which Spain exerted only relatively little influence.

As important as they can be, such material conditions have not been highly determining of Spain's actions. They are best thought of as permissive factors that have left considerable range for choice, and not only for Spain. Some countries with rich wind and solar resources, such as the United States, have been relatively slow to exploit them. Conversely, relatively limited renewable potential has not prevented other countries, such as Germany, from heavily promoting renewable power.

### *International Social Structures*

Thus somewhat more determinative of policy preferences, at least in Spain's case, has been the international social structures in which the country has been embedded, and by far the most important of these with regard to renewable power has been its membership in the European Union (EU). As we have seen, the EU has produced a series of renewable energy targets that members have been expected to meet, individually or collectively. A 1997 White Paper established a goal of providing 12% of primary energy consumption from renewable sources by 2010. A 2001 directive set a target of 29.4% of electricity from renewables by 2010. A 2009 directive required Spain to generate 20% of its final energy consumption from renewable sources by 2020. And, most recently, a 2018 directive established an ambitious collective renewable energy target for 2030 of at least 32%.

At the same time, EU membership has constrained the policy mechanisms Spain could employ to meet its targets. A notable example has been EU rules about state aid. These contributed to the shift from guaranteed levels of support for all plants that qualified to the use of auctions as a way of promoting renewable power more cost-effectively, although Spain's understandable desire to reduce the cost of support also played an important role in this shift.

To be sure, EU directives and regulations are also somewhat endogenous to the policy process, insofar as Spain participates in their development. But Spain's ability to dictate these policies in any detail is highly limited, given the large number of member states and the influence of the supranational Commission. Perhaps more importantly, even this detailed guidance has left considerable room for setting national targets and designing policies to achieve them.

## Domestic Imperatives

A third set of factors that have fostered cross-party commonalities in policy preferences consists of broadly shared domestic imperatives, the relative importance of which have nevertheless shifted over time. Several of these—the desire to reduce foreign energy dependence and increase the security of energy supplies, concern about the environment, especially climate change, and expected socioeconomic benefits—have favored the promotion of renewable power. Others, primarily concerns about the costs of renewable power and the impact of high levels of support on the fiscal health of the country, have pushed in the opposite direction. Such factors have not been unique to Spain, but they have been powerful motivators nonetheless.

The security benefits of renewable energy were noted as early as the 1980 Energy Conservation Law, the first major energy legislation adopted under the 1978 constitution, which aimed to reduce Spain's dependence on external sources of hydrocarbons (Ley 82/1980). Since then, they have been a constant theme in Spain's successive renewable energy plans. Indeed, the importance of this motivation only grew in the 1980s, 1990s, and early 2000s, as Spain's dependence on fossil fuels from foreign sources steadily increased, both in absolute and relative terms, with rapidly rising energy consumption. Until the economic crisis of 2008, fossil fuels accounted for roughly 50% of electricity generation, and the share of coal, the one fossil fuel that had been produced domestically, had been steadily declining, especially after imported natural gas was introduced in the early 2000s (BP 2020). And as recently as 2017, fossil fuel imports still amounted to more than 73% of Spain's primary energy consumption (IDAE 1999, 1; 2005, 13, 332–334; 2011, xxxiv; MITECO 2020a, 60–63).

The environmental benefits of renewable energy gained prominence in the 1990s, especially after the adoption of the Kyoto Protocol, which committed Spain to limiting the growth of its greenhouse gas emissions. Along with energy conservation and energy efficiency, the substitution of renewable energy for fossil fuels, especially in the generation of electricity, has been regarded as a key tool for reducing emissions of greenhouse and other noxious gases and for limiting the environmental impact of the energy system more generally. Over time, the importance of decarbonizing the economy has only increased, to the point where it became the leading goal of Spain's 2020 energy and climate plan (IDAE 1999, 9–15; 2005, 8–9, 23–24; 2011, 654–660; MITECO 2020a, 11–14).

The potential socioeconomic benefits of renewable energy were also recognized by the 1990s, but the importance attributed to them has grown in tandem with actual—and estimates of potential future—renewable energy deployments. The specific benefits were manifold and included technological development leading to greater industrial competitiveness, the creation of new businesses and jobs, especially in rural and remote areas, and thus regional development, and overall economic growth. Indeed, the number of companies and jobs in the sector as well as its contribution to Spain's GDP grew steadily in the 1990s and 2000s. By 2011, the renewables sector was expected to support some 300,000 jobs and generate about 18 billion euros in wealth per year by 2020, with renewable power again contributing about two-thirds of the total. Implementation of the 2020 energy and climate plan was expected to have a comparable impact, while reducing income inequality, over the following decade (IDAE 1999, 16–25; 2005, 14, 25–27, 337; 2011, 627–633, 646–652; MITECO 2020a, 213–226).

Domestic imperatives have not always favored policies promoting renewable power, however. At times, these pro-renewable forces have faced and even been eclipsed by countervailing pressures. As examined in Chapter 4, during the late 2000s and early 2010s, competing concerns about the costs of renewable power, especially solar PV, and the impact of generous support schemes on the fiscal health of the country came to the fore. As a result, governments led by both parties prioritized policies that first limited growth in and then reduced support costs.

## Sources of Divergent Policy Preferences

Despite such commonalities, the principal parties and their respective governments have sometimes differed in terms of the degree of emphasis on promoting renewable power and the choice of means for doing so. Arguably, the Zapatero government put in place a more generous—and ultimately unsustainable—version of the special regime that had been set up by the preceding Aznar government, although this revision also reflected growing concerns about Spain's ability to reach its renewable targets for 2010. Likewise, the Rajoy government arguably took harsher measures to halt the growing cost of the special regime, including the moratorium on new support payments and a fundamental restructuring of how support levels were calculated, than had its predecessor, although this difference also reflected the increasing severity of Spain's financial distress. And most recently, the Sánchez government has arguably done much more to promote the revival of renewable power than did its predecessor, although this difference also reflects changes in the targets (2020 versus 2030) they have been shooting for.

And on at least one issue—self-consumption—the major parties have been sharply at odds with one another. The Rajoy government was slow to act to develop the necessary regulation and ultimately adopted measures that made investment in self-consumption highly unattractive. The PSOE and virtually all the other parties, in contrast, strongly advocated lifting those constraints, and the PSOE began the process of doing so almost immediately after forming a new government in 2018.

What have been the underlying causes of these differences in policy preferences? Again, we can identify at least three places to look: party ideology, variation in susceptibility to the influence of interest groups, and leader personality.

*Party Ideology*

Heywood (1995, 193) has noted the ideological maneuvering and moderation on the part of both major parties in Spain. Nevertheless, one might point to several aspects of party ideology that have tended to push party preferences in opposite directions when it comes to promoting renewable power. For example, the People's Party has generally taken a more conservative approach to fiscal and financial matters. This may explain the greater emphasis it assigned to staunching the flow of support costs attributable to the special regime. For its part, the PSOE has placed more emphasis on environmental protection and democratization of the energy system. From this would seem to follow its championing of self-consumption and, arguably, its greater willingness to support renewable power overall.

*Interest Groups*

A second potential source of cross-party differences in policy preferences has been variation in the influence of interest groups in general and specific interest groups in particular. The renewable power sector, and the electricity sector more generally, has been characterized by the presence of a growing number of interest groups. Initially, the key private sector and non-governmental actors were the major Spanish utilities, such as Endesa, Iberdrola, Naturgy (formerly Fenosa), and EDP España (formerly Hidrocantábrico). Traditionally, the companies have often hired former high-level government officials in the hopes of influencing government policies in their favor (Transparency International España 2014). The major utilities have also had a well-funded lobbying group (originally *Asociación Española de la Industria Eléctrica* or UNESA and now *Asociación de Empresas de Energía Eléctrica* or aelēc).

As the renewable power sector has grown, however, so has the number and potential influence of associated interest groups, including a variety of industry associations. The first to form, in 1987, was the Association of Renewable Energy Producers (*Asociación de Empresas de Energías Renovables* or APPA), which covered all forms of renewable energy. Since then, it has been joined by more specialized associations representing wind power (*Asociación Empresarial Eeólica* or AEE), solar photovoltaic (*Unión Española Fotovoltaica* or UNEF and, before that, *Asociación de la Industria Fotovoltaica* or ASIF), and even concentrated solar power (*Asociación Española para la Promoción de la Industria Termosolar* or PROTERMOSOLAR). Also prominent has been an organization joining many of the tens of thousands of small investors in solar PV plants in the 2000s, who were hit particularly hard by the measures adopted to slow the growth of and then to reduce the tariff deficit (*Asociación Nacional de Productores de Energía Fotovoltaica* or ANPIER). Finally, mention should be made of the non-profit and non-governmental organizations (NGO) working in this area, including the *Fundación Renovables* and the *Real Instituto Elcano* as well as many of the usual environmental NGOs, such as Friends of the Earth, Greenpeace, and the World Wildlife Foundation.

All of these interest groups have sought to be heard and to influence policy in varying ways and to varying degrees. Many have issued press releases, sponsored conferences, and published reports for public consumption, and on occasion, they have even organized public protests. They have also taken advantage of formal channels to influence policy directly, including the provision of testimony before the Congress and government commissions, the submission of comments (*alegaciones*) during the periods of public review usually required of proposed legislation and regulation, and formal meetings with government officials. Prominent examples from the preceding chapters include the processes leading up to the adoption of *Real Decreto* 661/2007 and the development of the Integrated National Energy and Climate Plan (*Plan Nacional Integrado de Energía y Clima* or PNIEC) from 2018 through 2020. Presumably, such public activity has been accompanied by behind-the-scenes efforts to develop and exploit informal channels of influence.

Much harder to identify and evaluate is what impact interest groups have actually had on the policies pursued by different governments. On a few occasions, some have claimed that their proposals were adopted. More often, they have expressed disappointment or dissatisfaction with the results. And just as often, they have not commented on specific outcomes. It has been alleged, as noted in Chapter 4, that the major utilities had the ear of officials in the Rajoy government, while the renewable sector was largely ignored. For their part, the Sánchez governments have seemingly been more receptive to the views of the actors in the renewable power sector. But as a general rule, without speaking with the people involved, it is difficult, if not impossible, to identify direct links between influence attempts and the details of policies.

*Leader Personalities*

Given the nature of the policymaking process in Spain, with so much authority concentrated in the hands of such a small number of people, one might expect individual personalities to play an outsized role in determining policy outcomes. Without access to internal deliberations or conducting confidential interviews, however, evidence of such influence is particularly difficult to surface. The strongest claims in this regard unearthed in research for this book concern the roles of Alberto Nadal and Álvaro Nadal in determining the Rajoy governments' policies on self-consumption, as noted in Chapter 6. Also noteworthy was the decision by Teresa Ribera, the Minister of Ecological Transition, to raise the overall renewable energy target in the PNIEC from 40 to 42% at the last minute (Roca 2019). As a general rule, however, the roles that particular individuals may have played in shaping policy have not been given prominence.

## OTHER NATIONAL ACTORS

An exploration of national level influences on Spanish policy would be incomplete without considering several other actors, beginning with the independent regulators of the energy sector: the National Energy Commission (*Comisión Nacional de Energía* or CNE) and its successor, the National Commission on Markets and Competition (*Comisión Nacional de los Mercados y la Competencia* or CNMC). The CNE was established in 1998 to oversee and regulate the electricity and hydrocarbon markets as they were opened up to competition. At that time, it subsumed the pre-existing National Commission for the Electrical System, which had been created just the year before in the Electricity Sector Law (Ley 54/1997; Ley 34/1998; see also Real Decreto 1339/1999). Then, in 2013, the CNE was combined with the national competition authority and six other regulatory bodies in disparate areas to form the comprehensive CNMC (Ley 3/2013; see also del Río 2016, 21–22).

An important function of the CNE and the CNMC (henceforth just CNMC) has been to act as a consultative organ for the government on energy matters and, in particular, to participate in the process of elaborating laws, regulations, plans, and tariffs in the energy sector. In practice, this has meant that the body has reviewed and prepared a written report, typically including recommendations, on each proposed law, decree-law (*Real Decreto-ley*), and regulation (*Real Decreto*) developed by the executive. The CNMC's recommendations have not been binding on the government, but they have typically carried considerable weight, given the body's independence, and have often included helpful advice. A recent example of this advisory function can be seen in the development of the Law on Climate Change and the Energy Transition (LCCTE). In 2020, the CNMC recommended that the provisions in a draft be re-formulated as amendments to the 2013 Electricity Sector Law, a change that the Sánchez government subsequently adopted.

On other occasions, however, governments have simply ignored the body's recommendations, as exemplified by the debate over self-consumption. The CNE criticized the proposed backup tax (*peaje de respaldo*), characterizing it as discriminatory, and called for its removal. But the Rajoy government stood firm and included the tax in the final regulation.

It should be noted that the CNMC provides another avenue for influence by private actors. It is supported by a Consultative Council for Electricity, which includes many of the interest groups noted above. As a part of the process of preparing its reports, the CNMC typically solicits written comments from the members, and it sometimes holds hearings for the membership.

The importance of the CNMC was increased in early 2019, when a decree-law expanded its competencies in order to bring them into line with the requirements of EU energy market directives (Real Decreto-ley 1/2019). Over the following months, the CNMC issued a number of regulatory "circulars," including one concerning the conditions of access and connection to the grid. The latter in particular set off a turf battle with the government that was only resolved the following year by a decision of the State Council (*Consejo de Estado*), the supreme consultative body of the Spanish state. The Council ruled that the CNMC's circular was necessarily subordinate to, and thus could only logically follow, a regulation adopted by the government (Monforte 2020).

Another national actor of note is the transmission system operator, *Red Eléctrica de España* (REE). Established in 1985, REE seeks to maintain the stability of the grid and the access of generators to it. In that capacity, it is responsible for preparing a medium-term planning document every six years and ensuring that the grid expansions and improvements called for in the documents are achieved. It also is a member of the CNMC's Consultative Council for Electricity, where it has an opportunity to express views on legislation and regulation proposed by the government. And since mid-2019, it has played a key role in addressing the speculative bubble in access permits by rejecting applications for additional generation capacities that could not be accommodated by the grid. By the end of 2020, it had denied access permits concerning some 120 GW of capacity, or nearly three-quarters of the amount that had already been approved.

## The Regions and Municipalities

The politics of renewable power in Spain have not been confined to the national level. Another potentially significant set of actors consists of the regional and municipal governments. Spain is divided into 17 autonomous communities (*comunidades autónomas* or CCAA) and contains thousands of municipalities, including two autonomous cities. The regional authorities in particular have played noteworthy roles in the development of renewable power (IEA 2009, 94; Linares and Labandeira 2013, 4–5). Indeed, an entire book could be written about their many activities in this policy area.

### The Constitutional and Legal Basis of Authority

When it comes to the distribution of legal authority across levels of governance, Spain is somewhat of a paradox. On the one hand, it is not a federal but a unitary country, with sovereignty vested in the state as a whole, represented by the central institutions of government. On the other hand, Spain is highly decentralized. The state has, to varying degrees, devolved power to the regions, which enjoy limited autonomy and self-government. Indeed, the degree of decentralization has been characterized as greater than that in some federal countries and perhaps the greatest among all unitary states (del Guayo Castiella 2014, 55).

This complicated set of relationships is regulated by the constitution and a series of autonomy statutes that have been negotiated between the central government and the regions over the years. According to del Guayo Castiella (2014, 56–57), many state powers are shared in some way with the regions. Conversely, many regional powers are not exclusive, but to be exercised within the general framework laid down by the state. As a result, there is wide scope for conflict.

### Motivation

Regional and municipal governments have been motivated to promote renewable power for many of the same reasons that the central government has done so. The former have regarded the construction and operation of renewable power installations as valuable sources of development, employment, and wealth. Nor have regional and local governments taken a passive approach to securing these benefits. In some cases, the regions have required developers to build manufacturing or assembly facilities in order to obtain the necessary permissions and licenses (Linares and Labandiera 2013, 4–5). And prior to the advent of national auctions, they increasingly used bidding procedures for allocating authorizations in ways that maximized the benefits provided by project developers (Iglesias et al. 2011).

### Renewable Power Planning and Promotion

The regions have also been actively engaged in renewable energy planning to one degree or another for decades. By the late 1990s, they had all established renewable power targets. Indeed, the targets in the first national renewable energy plan of 1999 were simply a compilation of those set at the regional level (IDAE 1999, 77). And for years thereafter, regional plans and strategies were updated on a regular basis and taken into account in national renewable energy plans (see, e.g., IDAE 2005, 61; 2010, 113 and appendix).

These regional plans have frequently been accompanied by promotional efforts, which have taken many forms, including subsidies, tax deductions, incentives, and simplified administrative procedures, and are too numerous to list. Table 8.1 presents the more important measures adopted by one region, Andalusia, alone. Most recently, the region's Energy Strategy 2020 proposed no less than 117 distinct actions, many of which were intended to promote renewable energy. The region is one of the more active in this area but far from unique. Linares and Labandiera have shown how variations in these promotional efforts have impacted the distribution of wind power plants around the country (2013, 7–9).

Beginning in the mid-2010s, the regions became particularly active in promoting self-consumption and small-scale PV systems, especially during the time that the PP's restrictive regulations were in force. Prior to the lifting of those constraints by the Sánchez government, regional financial

**Table 8.1** Energy plans and promotional measures in Andalusia

Energy Plan of Andalusia 1995–1999 (PLEAN)
Renewable Energy Promotion Plan (PROSOL) 1995–1999
Order of September 30, 2002 (prioritized grid access and connection by renewable plants)
Energy Plan of Andalusia 2003–2006 (PLEAN)
Order of July 18, 2005 (established incentive program for sustainable energy)
Law 2/2007 (March 27) on the Promotion of Renewable Energy and Energy Saving and Efficiency
Sustainable Energy Plan of Andalusia 2007–2013 (PASENER)
Decree 169/2011 (May 31) (approved regulation to promote renewable energy)
Energy Strategy of Andalusia 2020 (2015)
Energy Strategy Action Plan 2016–2017
Decree-Law 7/2018 (June 26) (simplified rules for the development of renewable energy)
Incentive Program for Sustainable Energy Development in Andalusia through 2020

support in various forms totaled well over 1 billion euros (UNEF 2017, 40–43 and 52; 2018, 6–7).

### Authorization of Renewable Power Plants

Perhaps the most salient role of the regions has been in the authorization of renewable power plants. Under Article 149 of the Constitution, an exclusive competence of the central government is the authorization of electrical installations whenever their use affects multiple regions or their energy is transported across regional boundaries. In principle, however, this provision has meant that the regions have exercised at least some authority over smaller installations that have no extra-regional ramifications. With regard to renewable power, the potential ambiguity of this relationship was addressed as early as the 1997 Electricity Sector Law, which granted the regions the power to issue the necessary approvals for installations in the special regime, which covered renewable power plants of up to 50 MW (Ley 54/1997, Articles 27 and 28).

As a general rule, this power has fostered the deployment of renewable power, as regional governments have been interested in building as much capacity as possible (Linares and Labandiera 2013, 6; Iglesias et al. 2011). To that end, the regions have frequently simplified and shortened the administrative processes required to commission new power plants. On a few occasions, however, the possession of this authority has worked against deployment, as when a handful of regions imposed a moratorium on authorizations in the mid-2000s (IDAE 2005, 46, 58).

A related function has been the registration of plants that have been completed and commenced operation. *Real Decreto* 661/2007 (see Chapter 3) provided for initial registration at the regional level, at which point the plant would qualify for support. This arrangement contributed to the subsequent bubble in PV deployment, as the central government was not able to track just how much support had been awarded and thus respond in a timely manner. Subsequently, the central government, under both PSOE and PP leadership, sought to reassert exclusive authority for the registration of new plants in the special regime as part of its efforts to limit the growth of support costs. Nevertheless, a May 2018 ruling of the Constitutional Court granted the regions the right to regulate the registration of self-consumption facilities, a right that the Rajoy government had sought to reserve for the central government in *Real Decreto* 900/2015 (UNEF 2018, 64–65).

Another source of influence has been the exclusive regional competencies with regard to land-use planning. This power has also cut both ways. In Andalusia, for example, lax planning requirements have facilitated the construction of new plants on agricultural land. In the case of Catalonia, however, restrictive landscape protections throttled the deployment of grid-scale renewable installations for a decade (del Guayo Castiella 2014, 64; Prados 2010, 6905 and 6907; Decreto 147/2009).

Finally, the regional governments have been able to behave much like interest groups at the national level, seeking to influence the policies of the central government. They serve on the CNMC's Consultative Council for Electricity. They are consulted by REE as it develops its six-year investment plans. They are free to comment on proposed national laws and regulations. And under some governments, their energy ministers have met regularly with the corresponding national minister.

*Municipalities*

We conclude this section with a word about Spain's many municipalities. Overall, they have much more limited powers than do the regions, but their resources have nevertheless been consequential. In particular, since at least the early 2000s, they have been authorized to offer tax deductions for renewable power projects. In recent years, these incentives have been focused on self-consumption facilities. A 2018 study of Spain's 77 largest cities found that 45 offered deductions, typically up to the maximum of 50%, in the real estate tax that they collected (*Impuesto sobre Bienes Inmuebles* or IBI), and some two-thirds offered a deduction of 50% or more on the tax on buildings, installations, and works (*Impuesto sobre Construcciones, Instalaciones y Obras* or ICIO) (Fundación Renovables 2018).

Overall, however, the regions and municipalities have remained relatively minor actors in the policy process. They have influenced the deployment of renewable power, but their impact has tended to be felt only on the margins. The primary drivers of renewable deployment have been the actions taken by the central government. This has been the case, moreover, whether the effect has been to allow and even to incentivize new renewable installations or to constrain them. This difference in impact has been due to differences in the inherent authority of the central government and in the resources to which it has had access. The central government has been able to invent new support systems and to devise generous means of financing them, when needed. And when it has put on the brakes, as it did with the moratorium in 2012 and the 2015 regulation on self-consumption, there was little the regions and municipalities could do to overcome them.

## Conclusion: The Future of Renewable Power Policymaking in Spain

This chapter has analyzed the political forces that have shaped support for renewable power in Spain over the last four decades. It has emphasized the role of the core executive in determining Spanish policy and identified factors that have influenced governments led by parties of different political stripes to pursue similar actions. It has also sought to account for those instances in which the policy preferences of different parties and their respective governments have diverged.

As prominent as the politics of renewable power in Spain have been so far, they are likely to become only more important in the future. Spain has recently set a goal of making its electrical system entirely renewable before 2050 (MITECO 2020b). Even if the target contained in the 2020 PNIEC of obtaining 74% of electricity from renewable sources by 2030 is reached, the last 26% will be the most difficult to achieve, given the intermittent nature of the most significant renewable sources, wind and solar. Doing so will likely involve a substantial, if not fundamental, transformation of the electric power system. Demand response, storage, interconnections, and other innovations are all likely to assume increasing prominence.

Will the policymaking process be up to the task? One can identify at least three areas in which important changes are occurring in the process and are likely to have an impact. One is fragmentation of the political spectrum at the national level. Beginning in 2015, *Podemos* (We Can), *Ciudadanos* (Citizens), and *Vox* have joined PP and PSOE as major vote-getters in general elections. Although PP and PSOE have remained the two most popular parties, they have consistently lacked parliamentary majorities since then and, as a result, have been forced to form unstable minority governments dependent on the forbearance of small partners in the Congress and, most recently, Spain's first coalition government, which itself failed to command a majority. The logical result of this new reality is more bargaining between and compromises among the major parties, which can only slow and complicate policymaking.

A second trend is the growth and maturation of interest groups with stakes in the renewable power sector. We have seen in previous chapters the proliferation over the years of relevant interest groups, whose potential influence has only increased in tandem with their contributions to the national economy. This trend should tend to help Spain maintain momentum as it strives to meet its renewable targets in 2030 and beyond, although interest groups will never have the influence they enjoy in, say, the United States because of stronger restrictions on lobbying and campaign finance in Spain.

A third factor will be possible shifts in the relative power and authority of the national and regional governments. Spain is awash in political forces favoring greater regional autonomy, if not outright self-determination, and the above-mentioned fragmentation of the national political space has given small regional parties unprecedented leverage. Most notably, the success of the efforts of PSOE and *Podemos* to form a governing coalition still depended on the abstention of the Republican Left of Catalonia (ERC) in the January 2020 vote on investiture. Regional parties may be able to exploit such opportunities in order to gain a greater role in shaping policy toward renewable power, should they wish.

Overall, however, it would be safe to expect that there will be no fundamental changes in the policy-making process. The core executive will almost certainly remain pivotal, even if it has to cope with a much more complicated environment characterized by the proliferation of national political parties, entrenched interest groups, and regional party influence.

## References

BP (British Petroleum). 2020. *Statistical Review of World Energy*. https://www.bp.com/en/global/corporate/energy-economics/statistical-review-of-world-energy.html.

Decreto 147/2009, de 22 de septiembre, por el que se regulan los procedimientos administrativos aplicables para la implantación de parques eólicos e instalaciones fotovoltaicas en Cataluña. *Noticias Jurídicas* (28 Sept. 2009). https://noticias.juridicas.com/base_datos/CCAA/ca-d147-2009.html.

del Guayo Castiella, Iñigo. 2014. "Promotion of Renewable Energy Sources by Regions: The Case of the Spanish Autonomous Communities." In *Renewable Energy Law in the EU: Perspectives on Bottom-up Approaches*, edited by Marjan Peeters and Thomas Schomerus, 53–74. Cheltenham, UK: Edward Elgar.

del Río, Pablo. 2016. "Implementation of Auctions for Renewable Energy Support in Spain." Report D7.1-ES (March). http://auresproject.eu/sites/aures.eu/files/media/documents/wp7_-_case_study_report_spain_1.pdf.

Fundación Renovables. 2018. *Análisis comparativo de bonificaciones fiscales al autoconsumo en las principales ciudades españolas* (June). https://fundacionrenovables.org/wp-content/uploads/2018/07/ANALISIS-COMPARATIVO-BONIFICACIONES-FISCALES-AL-AUTOCONSUMO-Subvenciones-IBI-ICIO.pdf.

Heywood, Paul M. 1995. *The Government and Politics of Spain*. New York: St. Martin's.

———. 1999. Power Diffusion or Concentration? In Search of the Spanish Policy Process. In *Politics and Policy in Democratic Spain: No Longer Different?*, edited by Paul M. Heywood, 103–123. London: Frank Cass.

IDAE (Instituto para la Diversificación y Ahorro de la Energía). 1999. *Plan de Fomento de las Energías Renovables en España* (Dec.). https://www.idae.es/uploads/documentos/documentos_4044_PFER2000-10_1999_1cd4b316.pdf.

———. 2005. *Plan de Energías Renovables en España (PER) 2005–2010* (Aug.). https://www.idae.es/publicaciones/plan-de-energias-renovables-en-espana-2005-2010.

———. 2010. *Plan de Acción Nacional de Energías Renovables de España (PANER) 2011–2020* (30 June). https://www.mincotur.gob.es/energia/desarrollo/EnergiaRenovable/Documents/20100630_PANER_Espanaversion_final.pdf.

———. 2011. *Plan de Energías Renovables (PER) 2011–20* (11 Nov.). http://www.idae.es/file/9712/download?token=6MoeBdCb.

IEA (International Energy Agency). 2009. *Energy Policies of IEA Countries: Spain 2009 Review*. Paris: OECD.

Iglesias, Guillermo, Pablo del Río, and Jesús Ángel Dopico. 2011. Policy analysis of authorization procedures for wind energy deployment in Spain. *Energy Policy* 39: 4067–4076.

Ley 54/1997, de 27 de noviembre, del Sector Eléctrico (28 Nov.). https://www.boe.es/eli/es/l/1997/11/27/54.

Ley 34/1998, de 7 de octubre, del sector de hidrocarburos (9 Oct.). https://www.boe.es/eli/es/l/1998/10/07/34/con.

Ley 3/2013, de 4 de junio, de creación de la Comisión Nacional de los Mercados y la Competencia (6 June). https://www.boe.es/eli/es/l/2013/06/04/3/con.

Linares, Pedro, and Xavier Labandeira. 2013. "Renewable Electricity Support in Spain: A Natural Policy Experiment". *Economics for Energy WP 04/2013*. https://repositorio.comillas.edu/rest/bitstreams/18834/retrieve.

MITECO (Ministerio para la Transición Ecológica y Reto Demográfico). 2020a. "Borrador Actualizado del Plan Nacional Integrado de Energía y Clima 2021–2030" (20 Jan.). https://www.miteco.gob.es/images/es/pnieccomp leto_tcm30-508410.pdf.

———. 2020b. *Estrategia de Decarbonización a Largo Plazo 2050: Estrategia a Largo Plazo para Una Economía Española Moderna, Competitiva y Climáticamente Neutra en 2050.* (Nov.). https://www.miteco.gob.es/es/prensa/doc umentoelp_tcm30-516109.pdf.

Monforte, Carmen. 2020. "La CNMC sale derrotada frente a Ribera en la pugna por las conexiones eléctricas." *CincoDías* (2 July). https://cincodias.elpais. com/cincodias/2020/07/02/companias/1593669707_740312.html.

Prados, María-José. 2010. "Renewable Energy Policy and Landscape Management in Andalusia, Spain: The Facts." *Energy Policy* 38: 6900–6909.

Real Decreto 1339/1999, de 31 de julio, por el que se aprueba el Reglamento de la Comisión Nacional de Energía (24 Aug.). https://www.boe.es/eli/es/ rd/1999/07/31/1339.

Real Decreto-ley 1/2019, de 11 de enero, de medidas urgentes para adecuar las competencias de la Comisión Nacional de los Mercados y la Competencia a las exigencias derivadas del derecho comunitario en relación a las Directivas 2009/72/CE y 2009/73/CE del Parlamento Europeo y del Consejo, de 13 de julio de 2009, sobre normas comunes para el mercado interior de la electricidad y del gas natural (12 Jan.). https://www.boe.es/eli/es/rdl/ 2019/01/11/1.

Roca, Ramón. 2019. "La ministra Ribera decidió a última hora y por sorpresa aumentar hasta el 42% el objetivo de renovables en el PNIEC." *El Periódico de la Energía* (8 March). https://elperiodicodelaenergia.com/la-ministra-rib era-decidio-a-ultima-hora-y-por-sorpresa-aumentar-hasta-el-42-el-objetivo-de-renovables-en-el-pniec/.

Transparency International España. 2014. *An Evaluation of Lobbying in Spain: Analysis and Proposals.*

UNEF (Unión Española Fotovoltaica). 2017. *Informe Anual 2017.* https:// unef.es/wp-content/uploads/dlm_uploads/2017/07/informe-anual-unef-2017_web.pdf?utm_source=Emailing&utm_medium=Email&utm_campaign= Informe+anual+UNEF+2017.

———. 2018. *Informe Anual 2018.* https://unef.es/wp-content/uploads/dlm_ uploads/2018/09/memo_unef_2018.pdf.

# Index

**A**
Alliance for Self-Consumption, 115, 119
Anpier. *See* National Association of Photovoltaic Energy Producers
APPA. *See* Renewable Energy Business Association
ASIF. *See* Photovoltaic Industry Association
auctions, 143–147
  assessment of, 85–86
  January 2016, 82–84
  January 2021, 147
  July 2017, 85
  May 2017, 84–85
autonomous communities, 7, 45, 89, 115, 175–179
  Andalusia, 177, 179
  Aragon, 90
  Catalonia, 116, 179
  Extremadura, 90
  Murcia, 115
Aznar government, 21, 25–29, 32
Aznar, José Maria. *See* Aznar government

**B**
backup charge, 108–110
biomass, 36, 39, 46, 82
  capacity, 20, 30, 31, 36
  support levels, 31, 33, 42
  targets, 34

**C**
cannibalization, 142
capacity mechanisms, 152–153
cap and floor system, 41
Citizens party, 6, 180
  and self-consumption, 114–115, 117
climate change, 38, 136. *See also* Integrated National Energy and Climate Plan
CNC. *See* National Competition Commission

CNCM. *See* National Commission for Markets and Competition
CNE. *See* National Energy Commission
CO2 emissions, 18, 29, 39
coal, 14, 17, 152
Commission of Experts, 137
concentrated solar power, 43, 53
  capacity, 31, 46, 61
  support levels, 42
  targets, 39, 138
Congress of Deputies, 117, 118, 143, 155, 165
Constitutional Court, 116, 178
core executive, 6, 7, 155–156, 163–165
  powers, 165–166
  sources of policy preferences, 166–173
cost of renewable technologies, 86–87, 105, 121, 141, 167
Council of State, 111, 150, 175

**D**
decree-law, 165
del Río, Pablo, 4, 44, 91
demand management, 154
distributed generation, 58, 91, 111, 112. *See also* self-consumption

**E**
electricity consumption, 15, 38
electricity generation charge (2012), 65
electricity prices, 54, 142
electricity production, 13
Electricity Sector Law (1997), 27, 178
Electricity Sector Law (2013), 67, 82, 109, 143
electric utilities, 56, 108, 172

Energy Conservation Law (1980), 15, 16, 104
Energy Storage Strategy, 154
European Union, 5, 27, 66, 81, 168
  and Integrated National Energy and Climate Plan, 137, 138
  and interconnections, 155
  and power purchase agreements, 89
  and self-consumption, 117
  guidelines on state aid for energy, 81
  renewable energy directive (2009), 105
  renewable energy targets, 27, 33, 78, 136, 168

**F**
foreign energy dependence, 15, 26, 169
Franco, Francisco, 13

**G**
gas-fired combined cycle power plants, 56, 108, 152–153
González, Felipe. *See* González government
González government, 16, 18, 19, 21
grid access and connection, 38, 86, 87
grid access charge (2010), 65
grid parity, 87

**H**
hydroelectric power, 14, 15, 17, 18, 20, 36, 39
Hydrogen Roadmap, 153

**I**
Iberdrola, 108
ideology, 6, 123, 171

inspections of PV plants, 61–62
Integrated National Energy and Climate Plan, 137–138, 140, 152–154
interconnections, 154
interest groups, 7, 46, 56, 70, 123, 155, 171–173, 181
International Monetary Fund, 66
investor confidence, 68, 69, 140–141

## K
Kyoto Protocol, 26, 169

## L
Law on Climate Change and the Energy Transition, 143–145
LCCTE. See Law on Climate Change and the Energy Transition
legal challenges, 69, 116
Long-Term Decarbonization Strategy, 139

## M
market participation incentive, 35
merchant projects, 87
merit order effect, 142
Mir-Artigues, Pere, 4, 44

## N
Nadal, Alberto, 123
Nadal, Álvaro, 123
National Association of Photovoltaic Energy Producers, 70, 85, 116
National Commission for Markets and Competition, 82, 150, 151, 155, 173–175
   and Law on Climate Change and the Energy Transition, 174
   and self-consumption, 111, 119, 174
National Competition Commission, 110
National Energy Commission, 29, 40, 41, 57, 58, 62, 174
   and inspections, 62
   and self-consumption, 107, 109, 110
National Energy Plan (1983), 16
National Energy Plan (1991), 18, 104
National Plan for the Assignment of Emissions Rights (2004), 38
National Renewable Energies Action Plan (2010), 79, 105
National Storage Strategy, 153
net billing, 120
net metering, 106, 120
nuclear power, 14, 17, 110

## O
oil shock, 14, 15

## P
PEN. See National Energy Plan
People's Party, 6, 34, 106, 123, 166, 171, 180
PER. See Renewable Energies Plan
Photovoltaic Industry Association, 105
PNIEC. See Integrated National Energy and Climate Plan
*Podemos*, 6, 123, 144, 180
power purchase agreements, 88–89
President, 6, 164, 165
PSOE. See Spanish Socialist Workers' Party

## R
Rajoy government, 52, 78, 81, 114

and Integrated National Energy and Climate Plan, 137
and Law on Climate Change and the Energy Transition, 143
and self-consumption, 107–112, 114, 117
and tariff deficit, 57, 60, 66
auctions, 81–85
moratorium on new support, 60, 80
reductions in support levels, 64
Rajoy, Mariano. *See* Rajoy government
*Real Decreto* 244/2019, 118–120
*Real Decreto* 413/2014, 67
*Real Decreto* 436/2004, 32, 34–36
limitations, 37–38
support levels, 34–35, 37
*Real Decreto* 661/2007, 37, 41–43, 45, 178
impact, 43–44
support levels, 41–42
*Real Decreto* 900/2015, 112–113
*Real Decreto* 960/2020, 146
*Real Decreto* 1003/2010, 62
*Real Decreto* 1565/2010, 60, 63
*Real Decreto* 1578/2008, 58–59, 61
*Real Decreto* 1699/2011, 106
*Real Decreto* 2366/1994, 19, 20
*Real Decreto* 2818/1998, 28–29, 33
*Real Decreto-ley* 1/2012, 107
*Real Decreto-ley* 6/2009, 60
*Real Decreto-ley* 9/2013, 66, 108
*Real Decreto-ley* 14/2010, 63, 65
*Real Decreto-ley* 15/2018, 117–118, 150
*Real Decreto-ley* 23/2020, 145, 150, 153
reasonable profitability, 66
*Red Eléctrica de España*, 89, 148, 150, 175
regional governments. *See* autonomous communities
regulated costs, 55

Renewable Energies Plan (1986), 16
Renewable Energies Plan (1989), 17
Renewable Energies Plan (2005), 38–40
Renewable Energies Plan (2011), 79, 105
Renewable Energies Promotion Plan (1999), 29–31
renewable energy benefits, 169–170
Renewable Energy Business Association, 43, 84, 172
Renewable Energy Economic Regime, 145
Renewable Energy Foundation, 70, 115, 123
renewable power
capacity, 20
generation, 17, 20, 44, 68
potential, 167
support costs, 65, 68, 80
support levels, 33
targets, 29, 33, 78–80, 136, 138, 139
Ribera, Teresa, 143, 173

**S**

Sánchez government
and auctions, 89, 143, 143–147
and capacity mechanisms, 153
and grid access regulation, 152
and Integrated National Energy and Climate Plan, 138
and Law on Climate Change and the Energy Transition, 144
and rates of return, 141
and self-consumption, 117–120
and storage, 153–154
Sánchez, Pedro. *See* Sánchez government
self-consumption, 8, 101–124, 142
benefits, 102–103

capacity, 121
definition, 102
self-generation, 15, 104. *See also* self-consumption
self-production, 104. *See also* self-consumption
solar photovoltaic power, 19
    capacity, 17, 20, 31, 36, 43, 46, 61, 89, 90
    generation, 20, 44
    support costs, 53, 65
    support levels, 35, 42, 58–60, 63–64
    targets, 39, 138
Spanish Photovoltaic Union, 70, 85, 109, 123
Spanish Socialist Workers' Party, 6, 166, 171, 180
special regime, 25, 28
    origins, 19
    replacement, 66
    support costs, 32, 40, 52
specific remuneration regime
    criticism of, 66–68
storage, 153–154
sun tax. *See* backup charge
Supreme Court, 116, 118

**T**
tariff deficit, 51, 53, 65
    causes, 54–55

**U**
UNEF. *See* Spanish Photovoltaic Union
Union of the Democratic Center (UCD), 15

**V**
*Vox*, 6, 145, 180

**W**
wind power, 19, 21
    capacity, 17, 20, 31, 35, 46, 53, 61, 89, 90
    generation, 20, 44, 46
    support costs, 53
    support levels, 63
    targets, 30, 34, 39, 138

**Z**
Zapatero government, 26, 37, 40, 79
    and self-consumption, 106
    and tariff deficit, 57–59, 60, 65
    cuts in support levels, 63, 64
    renewable power targets, 78
Zapatero, José Luis. *See* Zapatero government

CPSIA information can be obtained
at www.ICGtesting.com
Printed in the USA
LVHW081910100222
710781LV00005B/311